Hydrogen Electrical Vehicles

Scrivener Publishing
100 Cummings Center, Suite 541J
Beverly, MA 01915-6106

Advances in Hydrogen Production and Storage

Publishers at Scrivener
Martin Scrivener (martin@scrivenerpublishing.com)
Phillip Carmical (pcarmical@scrivenerpublishing.com)

Hydrogen Electrical Vehicles

Edited by
Mehmet Sankır
Department of Materials Science and Engineering, TOBB University of Economics and Technology, Ankara, Turkey

and

Nurdan Sankir
Department of Materials Science and Engineering, TOBB University of Economics and Technology, Ankara, Turkey

Scrivener Publishing

WILEY

This edition first published 2023 by John Wiley & Sons, Inc., 111 River Street, Hoboken, NJ 07030, USA
and Scrivener Publishing LLC, 100 Cummings Center, Suite 541J, Beverly, MA 01915, USA

For more information about Scrivener publications please visit www.scrivenerpublishing.com.

Wiley Global Headquarters
111 River Street, Hoboken, NJ 07030, USA

For details of our global editorial offices, customer services, and more information about Wiley products visit us at www.wiley.com.

Library of Congress Cataloging-in-Publication Data

ISBN 978-1-394-16638-1

Cover image: Pixabay.Com
Cover design by Russell Richardson

Set in size of 11pt and Minion Pro by Manila Typesetting Company, Makati, Philippines

Printed in the USA

10 9 8 7 6 5 4 3 2 1

Contents

Preface

The decision of 28 countries to limit global warming to well below 2 degrees celsius in accordance with the Paris Agreement, can be realized by minimizing CO_2 emissions, which can only be accomplished by establishing a hydrogen ecosystem. A new geopolitical order is envisaged, in which sectors dealing with energy production, distribution, and storage, and thus an increasing carbon footprint, are reconstructed. In short, an economic order with new tax regulations is being created in which carbon footprints will be followed. This effort, which is called the "Green Deal," is defined in Europe as a new growth strategy aiming for net-zero CO_2 emissions. We know that transportation is responsible for about 24% of all CO_2 emissions. Therefore, any efforts to reduce emissions must include utilizing hydrogen in the transportation sector.

Fuel cells use hydrogen directly in most types of vehicles—from passenger cars to trains—without some of the disadvantages of batteries such as low energy density, high initial costs, and a slow charge. Therefore, the number of hydrogen filling stations required for fuel cells, about one-tenth of the number of fast-charging stations, can meet the same needs as batteries. Additionally, hydrogen charging is at least three times faster. Therefore, it is essential to emphasize that hydrogen-powered transportation is still the most reasonable way to reduce emissions.

As part of our "Advances in Hydrogen Production and Storage" series, this volume covers the cutting-edge technologies used in fuel cell-powered cars. Additionally, it highlights the research efforts presented in the literature while adding a valuable component to the area. It also discusses basic as well as advanced engineering details for both scientists and engineers in academia and industry.

There are seven chapters in the book. Chapter 1 introduces hydrogen and electrical vehicles. Hydrogen storage and compression systems are analyzed in Chapters 2 and 3, respectively. Chapter 4 discusses hydrogen propulsion systems for UAVs. The testing and evaluation of hydrogen fuel cell vehicles are covered in Chapter 5. Chapter 6 focuses on hydrogen production and polymer electrolyte membrane (PEM) fuel cells for electrical vehicles. The final chapter, Chapter 7, covers the issues concerning the power and durability of fuel cell vehicles.

In closing, we wish to thank the distinguished authors for their valuable contributions in reviewing the efforts made towards using hydrogen in electrical vehicles.

Editors
Dr. Mehmet Sankır and **Dr. Nurdan Demirci Sankır**
Department of Materials Science and Nanotechnology Engineering,
TOBB University of Economics and Technology

1

Hydrogen Electrical Vehicles

Ameen Uddin Ammar, Mohamad Hasan Aleinawi and Emre Erdem*

Sabanci University, Faculty of Science and Engineering, Materials Science and Nano Engineering, Orhanli, Istanbul, Turkey

Abstract
Hydrogen usage in electric vehicles is one of the domains which provide immense potential to explore in the race of energy efficient vehicles. As the world continuously moves towards clean and green environment, where all possible measures are been taken to minimize the carbon emission. Discussed here are the following: the numbers of hydrogen-utilized projects for energy, reducing prices of clean energy production facilities, and researching new methods to store hydrogen. Fuel cells which use hydrogen as the source of energy are proving themselves to be a serious candidate in this cause with numerous research are taking place to avail this opportunity.

***Keywords*:** Hydrogen electrical vehicles, zero carbon emission, hydrogen storage, green environment, hydrogen production, carbon capture, electrolysis, fuel cells

1.1 Hydrogen Usage in Electrical Vehicles

Probably the most crucial and challenging issue the world is facing nowadays is carbon emission. It has effects over several aspects, from environmental problems, power generation, economy, and many others. The world has reached the "no going back" point with carbon emissions, and we have to take effective measures in dealing with this challenge. Luckily, many nations and agencies started to intensify their efforts in order to face this problem. This can be sensed from International's Energy Agency (IEA) "net zero emission by 2050 roadmap" report [1]. IEA plans to make drastic changes to fuel sources and replacing fossil fuels which have a high carbon

**Corresponding author*: emre.erdem@sabanciuniv.edu

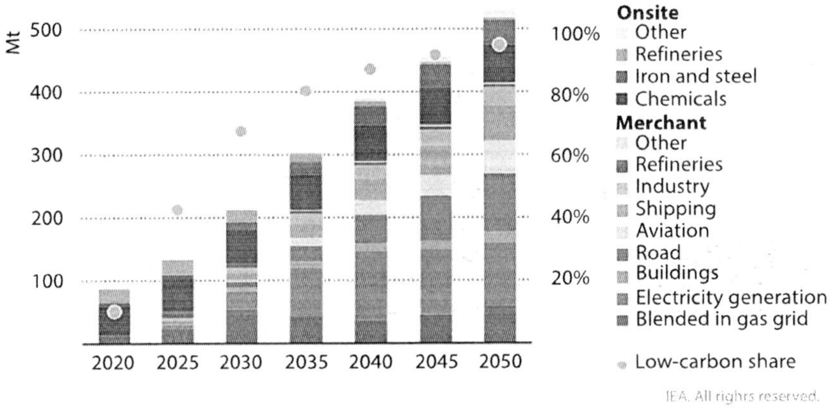

Figure 1.1 Hydrogen as a source of fuel future development plan. (Net Zero by 2050).

emission with hydrogen from pure hydrogen sources such as electrolysis or low carbon hydrogen sources such as methane and ammonia. Figure 1.1 above displays the future development plan for hydrogen usage as a fuel in different sectors.

We can see how the plan is to grow from below 100 Mt to greater than than 500 Mt between 2020 and 2050. We can also notice that using hydrogen as a fuel in transportation also plays a crucial rule in the upcoming years.

Fuel Cell Electric Vehicles (FCEV) uses energy extracted from hydrogen fuel cell as a source of fuel [2]. FCEV has zero carbon emissions, and the only emissions are water and heat. FCEV also has a higher efficiency compared to conventional Internal Combustion Engine (ICE) cars. Thanks to the regenerative braking system, a battery can be connected to brakes in order to store lost energy due to braking and use it elsewhere. FCEV has some advantages over all electric vehicles. First of all, FCEV takes its power from hydrogen tanks, which means it doesn't require long charging times like electric vehicles. Another clear advantage is that FCEV power is dependent on the hydrogen tank size, while electric vehicle's power depends on the size of the battery.

Figure 1.2 below shows the main components of FCEV [3]. A stack of fuel cells provides the required energy extracted from the hydrogen tank, and of course the output of this reaction is energy plus water without any carbon emissions. The battery at the back of the vehicle is connected to the braking system in order to harvest the lost energy due to braking and store it for future use.

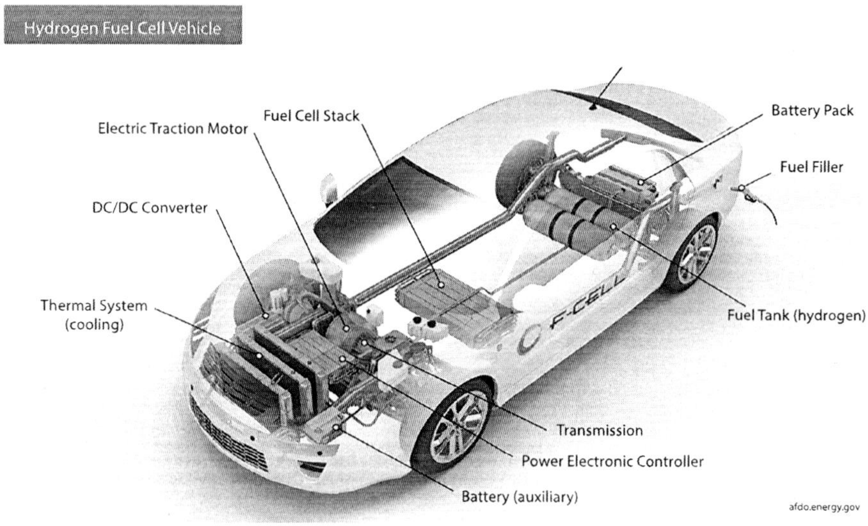

Figure 1.2 Components of FCEV (U.S. Department of Energy).

Bear in mind that FCEV is not only limited to cars. It can be used in busses, trains, or even as a source of power for space missions. The latter is actually very important since the output water can be of great use in space missions where there is no source of water. Figure 1.3 below shows the current and future development of Zero Emission Busses (ZEB) in California, USA.

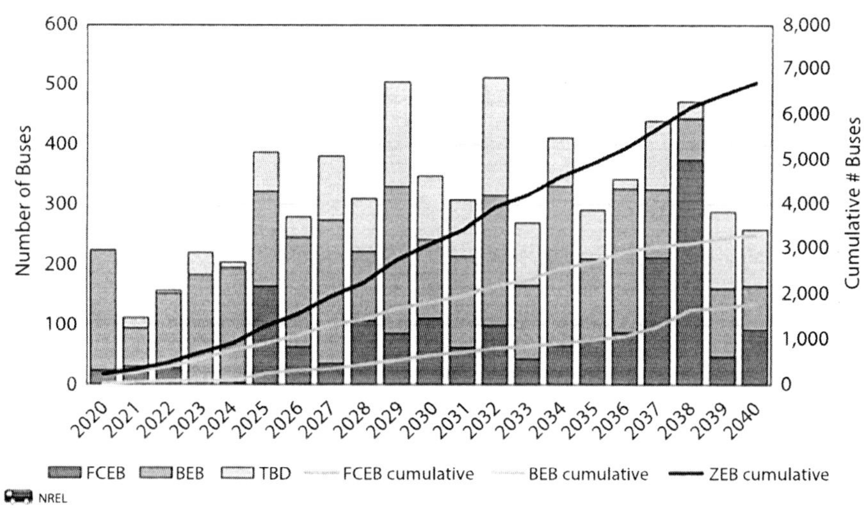

Figure 1.3 Current and future development plan for ZEB in California-USA (Fuel Cell Buses in U.S.).

Table 1.1 Current and future development plan for ZEB in regions other than USA (Fuel Cell Buses in U.S.).

Region	Number of ZEBs
Europe	1467
Asia	2518
Australia	100
South America	2
Total	4087

In the USA alone, the estimated number of ZEB between 2020 and 2040 is 7000, between Fuel Cell Electric Busses (FCEB) and Battery Electric Busses (BEB). Table 1.1 above shows the same development plan for regions other than the USA [4].

1.2 Hydrogen Production for Electrical Vehicles

Obviously, fuel cells need hydrogen in order to operate, and the process is 100% carbon free. However, the pure hydrogen sources so far haven't been carbon free. In fact, carbon emissions involved in hydrogen production from fossil fuels and coal are matched to and sometimes higher to carbon emissions from IC vehicles [5]. It is not enough to shift into hydrogen as a fuel to cut down carbon emissions. Hydrogen sources should also be carbon free in order to optimize this technology. Germany has proposed a 150 Million EUR project of 100 MW water electrolyzer that uses wind energy in order to produce pure hydrogen from water electrolysis. Producing hydrogen from renewable sources such as wind or solar energy is the key of success for this approach. Netherlands also is studying a 2 GW project similar to Germany. Austria is currently producing pure hydrogen from a 6 MW water electrolyzer. Japan, Canada, China, USA and many other countries are also investing in similar projects.

Figure 1.4 below shows the current status of water electrolysis projects.

An alternative and less costly solution is to use "Blue hydrogen production" which is producing pure hydrogen from fuel fossils with capturing CO_2 emissions. Of course, blue hydrogen is not CO_2 free. In fact, CO_2

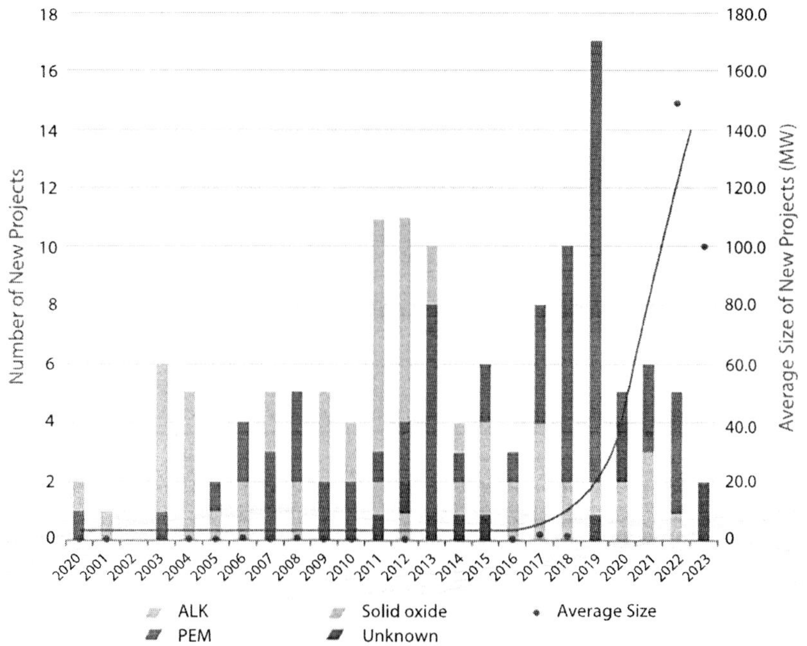

Figure 1.4 Current status of water electrolysis projects (Hydrogen: A Renewable Energy Perspective).

capture efficiency is estimated at 85–90% at best. However, it is still a good option to reduce carbon emissions and to provide pure hydrogen for use as a fuel. Figure 1.5 below shows the expected fall of price of electrolyzers between now and 2050.

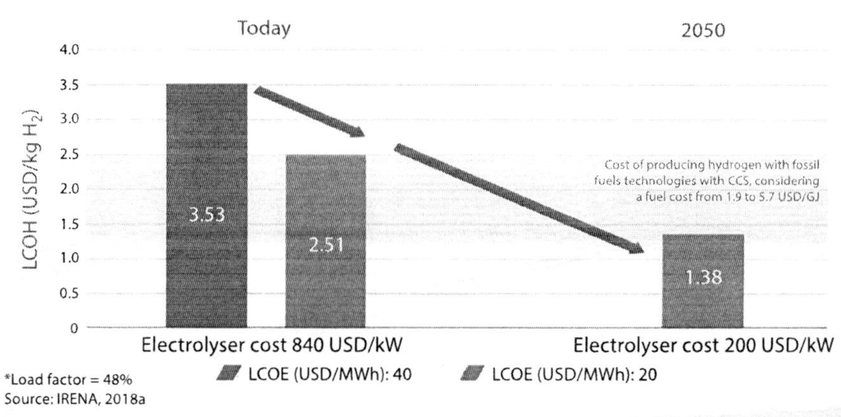

Figure 1.5 Prices of electrolyzers (Hydrogen: A Renewable Energy Perspective).

1.3 Hydrogen Storage Methods

Pure hydrogen can be stored in two main methods [6]:

1. Physical storage.
2. Chemical storage.

Physical storage implies that hydrogen atoms do not interact with the storing medium. It can be either compressed or liquefied cryogenically. Compressed hydrogen can be stored in gas cylinders, stationary storage systems such as underground reservoirs, glass microsphere, pipelines, and other methods.

Chemical storage on the other hand occurs when hydrogen atoms interact with the storage medium. It can be categorized as adsorption, such as adsorption by zeolites and metal-organic materials. It can be also categorized as absorption, such as metal hydrides. Finally, it can also be categorized as chemical interactions with storage materials such as liquid organic hydrides, ammonia and methanol, water reacting metals.

Each of the aforementioned methods has its pros and cons. Storing hydrogen physically is the simplest method. However, it requires a considerable amount of energy to compress the hydrogen or liquefy it. Chemical storage methods are effective, but still under development in order to optimize an efficient way to store pure hydrogen for usage as a fuel.

1.4 State-of-the-Art for Hydrogen Generation and Usage for Electrical Vehicles

Xu *et al.* [7] designed a liquid hydrogen storage tank to be used in remotely operated aircraft which will have increased endurance and will fly at high altitude. The work reported the basic structural design and analysis of the cryogenic liquid hydrogen tank, a thermal model was established for the tank and heat leakage of the support system was reduced by building insulating support. The result obtained shows the feasibility of the design and analysis method with stable structure and required mechanical strength.

Aceves *et al.* [8], showed that the cryogenic capable pressure vessels integrated within a cryo-compressed storage system can store high-density hydrogen, they have high thermal endurance compared to conventional liquid hydrogen tanks, and can improve the evaporative losses in automobiles. The developed system store hydrogen more efficiently, provide fast refueling, and are light in weight. The system developed was demonstrated on the hydrogen hybrid vehicle which enhanced the driving range on a single fuel tank with high thermal endurance.

Okumus *et al.* [9], developed a hydrogen generation system (HGS) using borohydride and a fuel cell system (FCS) to power and manufacture an unmanned aerial vehicle (UAV). The research works on preparing an economical and durable hydrogen generation catalyst through sodium borohydride solution by keeping it under high pressure in the reaction chamber and by controlling the flow rate of the fuel pump and heating device power. While for the fuel cell system of the UAV, a high-rate hydrogen generation system catalytic hydrolysis of $NaBH_4$ through transition metal catalysts was developed. The developed HGS and FCS generate 218 W power and show an energy density value of about 325 Wh/kg.

Ahluwalia *et al.* [10] performed the technical assessment of the onboard and off-board performance of cryo-compressed hydrogen storage tank which will be used in automotive applications. The on-board performance assessment includes weight, volume energetics, and refueling while the off-board assessment includes thermal management greenhouse gas emissions and energy efficiency, etc. The works show that a cryo-compressed storage system has the potential to meet the required target which has the appropriate gravimetric capacity, volumetric capacity, and in control hydrogen loss during dormancy under certain conditions of minimum daily driving.

Yamashita *et al.* [11], reported the manufacturing of a high-pressure hydrogen storage system which will be used in Toyota Mirai. The new hydrogen storage system used incorporated new components such as valves, regulators, and tanks which will provide increased hydrogen capacity without compromising on the interior space. The weight of the new storage system was reduced by using improved Carbon fiber reinforced plastic (CFRP) and refueling performance was also improved on the developed hydrogen storage system by ensuring compatibility with the SAE J2601 and J2799 standards for communication between the hydrogen station and vehicle.

Zhang *et al.* [12], presented a thorough system design and control strategy of the vehicles that are utilizing hydrogen energy. The work presented hydrogen supply, hydrogen storage method, safety protocols of the hydrogen vehicle system. In the hydrogen vehicle, three different types of the fuel storage system are used for a brief period, that are high pressure, liquid storage, and metal oxide storage system to see compare performance. Proton exchange membrane fuel cell (PEMFC) is used as fuel cells and the driving form and intelligent control of the PEMFC hybrid vehicle are analyzed.

Gany *et al.* [13], showed the benefits of utilizing electric power and using onboard hydrogen generation storage for marine vehicles. Aluminum–water reaction is carried out for the generation of hydrogen and electric energy vehicles, the method used shows a high reaction rate and increased hydrogen production at room temperature. The use of this storage system with PEM fuel cell provides a compact method for electrical energy storage which was feasible for long duration and long-range. A model boat equipped with a hydrogen reactor, fuel cell, and electric motor has been constructed and operated, demonstrating the technology.

Ananthachar *et al.* [14], showed the comparison of energy efficiencies of different types of the hydrogen storage system used in a fuel cell vehicle, the work also analyzed the reformer system in a fuel cell vehicle. Three of the most used fuel storage methods on fuel cell vehicle were compared which includes (a) compressed hydrogen gas storage, (b) metal hydride storage, and (c) onboard methanol-reformer system. The compressed hydrogen gas stored fuel cell vehicle was concluded to be the most energy-efficient vehicle. The compressed hydrogen gas tank vehicle storage at 33% is either slightly above or equal to the battery-electric car depending on the source of fuel in the power plant producing electricity for the battery charging. The compressed gas system is simple in design; lighter in weight compared to the other system, and is far more energy-efficient.

Deluchi *et al.* [15], showed an analysis of the performance, technology, safety, and environmental aspects of the solar-hydrogen fuel cell vehicles, a fuel cycle where hydrogen is generated by the solar-electrolysis of water and after that used in a fuel-cell-powered electric motor vehicle. The developed vehicle will produce very little pollution. Hydrogen fuel cell vehicle shows the best feature from both battery-powered and fossil fuel vehicles, which includes zero-emission, quiet operation, high efficiency, fast refueling time, and long life with long-range.

1.5 Conclusions

In short, in the year of 2022 it is possible to say that Hydrogen is the next wave for electrical vehicles due to its numerous advantages. Therefore in very near future Hydrogen will play the key role in the mobility which uses renewable energy systems. Recently the fuel cell research and its industrial applications gained enormous momentum and one of the best ways to use the hydrogen as energy source is to build smart fuel cell systems having high energy efficiency and zero CO_2 emission. Using of such systems will serve the decarbonization both in industry and transport. In addition, it is envisaged that in near future fuel cells may complement batteries and supercapacitors to decarbonize energy storage systems. By this in case of transportation it is possible to predict that in about 50 years zero emission vehicles will be mobile on the roads. Further Hydrogen busses and trucks, and fuel cell trains (hydrail) are next applicable vehicles having heavy loads and long distances.

References

1. IEA, Net zero by 2050 – A roadmap for the global energy sector, pp. 1–224, 2021.
2. Briguglio, N., Andaloro, L., Ferraro, M., Antonucci, V., Fuel cell hybrid electric vehicles. *Intechopen*, 1–23, 2011.
3. Council, N., Sciences, D., Systems, B., Economy, C., Assessment of fuel economy technologies for light-duty vehicles, pp. 1–218, 2011.
4. Eudy, L. and Post, M., Fuel cell buses in U.S. transit fleets: Current status 2020. *Nat. Renew. Energy Lab.*, 1–57, 2020.
5. *2nd Hydrogen Energy Ministerial Meeting in Tokyo, J., Hydrogen: A renewable energy perspective*, 2019.
6. Yartys, V.A. and Lototsky, M.V., *An overview of hydrogen storage methods*. Springerlink, 2004.
7. Xu, W., Li, Q., Huang, M., Design and analysis of liquid hydrogen storage tank for high-altitude long-endurance remotely-operated aircraft. *Int. J. Hydrog. Energy*, **40**, 46, 16578–16586, 2015.
8. Aceves, S.M. *et al.*, High-density automotive hydrogen storage with cryogenic capable pressure vessels. *Int. J. Hydrog. Energy*, **35**, 3, 1219–1226, 2010.
9. Okumus, E. *et al.*, Development of boron-based hydrogen and fuel cell system for small unmanned aerial vehicle. *Int. J. Hydrog. Energy*, **42**, 4, 2691–2697, 2017.
10. Ahluwalia, R.K. *et al.*, Technical assessment of cryo-compressed hydrogen storage tank systems for automotive applications. *Int. J. Hydrog. Energy*, **35**, 9, 4171–4184, 2010.
11. Yamashita, A. *et al.*, *Development of high-pressure hydrogen storage system for the Toyota "Mirai"*, SAE International, SAE 2015 World Congress & Exhibition, United States, 2015.
12. Zhang, Z. and Hu, C., System design and control strategy of the vehicles using hydrogen energy. *Int. J. Hydrog. Energy*, **39**, 24, 12973–12979, 2014.
13. Gany, A., Elitzur, S., Rosenband, V., Compact electric energy storage for marine vehicles using on-board hydrogen production. *J. Ship. Ocean Eng.*, **5**, 4, 151–158, 2015.

14. Ananthachar, V. and Duffy, J.J., Efficiencies of hydrogen storage systems onboard fuel cell vehicles. *Solar Energy*, **78**, 5, 687–694, 2005.
15. DeLuchi, M.A. and Ogden, J.M., Solar-hydrogen fuel-cell vehicles. *Transp. Res. Part A: Policy Pract.*, **27**, 3, 255–275, 1993.

2

Study on a New Hydrogen Storage System - Performance, Permeation, and Filling/Refilling

Leonardo Ribeiro[1,2], Gustavo F. Pinto[1,2*], Andresa Baptista[1,2] and Joaquim Monteiro[1,2]

[1]*ISEP—School of Engineering, Polytechnic of Porto, Porto, Portugal*
[2]*INEGI—Instituto de Ciência e Inovação em Engenharia Mecânica e Engenharia Industrial, Porto, Portugal*

Abstract
Hydrogen is increasingly a possibility to replace fossil fuels in vehicle propulsion. However, due to the properties of H_2, there are still unsolved difficulties for such replacement. This chapter presents a new way for H_2 storage in vehicles, propelled either by reciprocating engines or by stacks of fuel cells – H_2 is at high pressure within small spheres randomly packed in conventional vehicle's tanks. The H_2 supply by the spheres to the consumer or the refilling of such spheres is controlled by a chip. The purpose of this study is to evaluate the performance of the new system, comparing it with conventional fuel storage systems through parameters such as the energy stored by volume (VED) and the gravimetric energy density (GED); moreover, to evaluate the H_2 leaks by permeation and see which is the best set of materials for each part of the system; to check the compliance of the system with safety standards; finally, to propose a plausible method for the filling/refilling of spheres. It was concluded that GED is threefold and VED is about half the homologous values of conventional systems. To guarantee safety, the spheres were built with two concentric layers, aluminum for the liner and CFEP for the structural layer, Si for the chip, and aluminum for the tank.

Keywords: Hydrogen, permeation, storage safety, energy systems, green energy, vehicle propulsion, packing factor, envelope tank

**Corresponding author*: gflp@isep.ipp.pt

2.1 Introduction

Around 87% of worldwide man-generated CO_2 emissions are due to fossil fuel combustion [1]. In the last few years, governments of several countries have shown great interest to implement an effective worldwide plan to halt the global warming of the Earth and tackle the challenges linked to climate change [2]. This effort was patent in the Paris Agreement, which aims to decarbonize the world economies [3]. However, the goals imposed by the Paris Agreement jeopardize the reduction of energy poverty and, therefore, have been questioned [4].

In the path for decarbonization, it is foreseen that hydrogen (H_2) will be considered an important energy vector, either for energy storage or for propulsion [2, 5, 7]; some experts foresee that, in the future, H_2 will be used as a general-purpose energy vector, either for energy storing or transport [5, 6]. The most economically powerful countries like, for example, China, [8–10], Germany [11], Japan [12] and the United States [13], are getting ready to boost the consumption of H_2, which is shown by their leader's preparedness to build a friendly economic and political background.

Energy solutions are seen as the way forward for a sustainable future; several authors have started to study earnestly this subject [4, 14, 15]. Dincer and Acar [14] claim that smart energy is associated with solutions involving H_2. These authors highlight the importance of a carbon-free economy, where sustainable H_2 production is still under development. They stated, as well, that the minimum and necessary requirements for an intelligent solution based on H_2 towards a sustainable future are: (i) use of renewable energy and its control, (ii) energy conservation, (iii) clean energy, (iv) the use of smart grids and (v) storage of energy carriers and more efficient chemical species [14].

The use of H_2 as fuel is at an early stage. To go forward, it is necessary to create good infrastructures for storage as well as distribution. Refueling stations must be safe and allow fast refilling to meet the basic and necessary conditions for common use [16, 17].

Soon, for the propulsion of vehicles, stacks of Proton Exchange Membranes Fuel Cells (PEMFC), could be a competitive and ecological alternative to batteries. The H_2 for fuel cells can achieve large-scale use in vehicles as renewable energy becomes widespread [18–20]. One of the main challenges to get this is the improvement of H_2 filling stations, in terms of safety and quick refilling [21].

There is also a rising interest in the use of H_2 for vehicle propulsion through combustion engines.

The storage of H_2 in a vehicle is recognized as of great importance: nearly 70% of publications are about the storage of H_2 [22]. In fact, in consequence of the growing interest in the use of H_2 for vehicle propulsion, the interest for storing this fuel in vehicles has also risen.

There are essentially two ways to store H_2 [6]. The first way is to store it as (i) compressed gas, (ii) cryogenic liquid, (iii) adsorbed on carbon nanofibers, or (iv) adsorbed on a metal, such as a reversible metal hydride [6, 23]. In the second form, hydrogen is combined with certain chemical species to obtain methanol (CH_3OH), ammonia (NH_3), etc.

The most common storage method for H_2, in vehicles or in stationary applications (filling stations, underground caves, and so on...) are the man-made pressurized containers of various shapes and sizes. Larminie *et al.* [23] focus on some advantages over H_2 storage compared to other storage systems. They emphasize CGH_2, because of its unlimited storage time, simplicity, and no purity requirements.

Other studies have been carried out, such as the environmental impact of using H_2 as a fuel in vehicles, where H_2 powered vehicles are compared with electric vehicles charged with electricity generated by renewable sources, or green energy (sun, wind, tides, among others) [6, 24]. In these research works, reference is made to the great interest in the issue of H_2 storage. They claim that it is economically viable to store large amounts of energy, in the form of H_2, considering the seasonal availability of green energy [25–28]. In fact, it is possible to use an eventual surplus of renewable energy production to obtain hydrogen that will be stored until it is required by consumers [6, 22–28]. In the case of such surpluses, Tarkowski [29] studied the feasibility of underground hydrogen storage using salt caverns, deep aquifers, depleted gas fields and depleted oil fields.

It is well known that because H_2 has a very low density, to introduce a large mass of gaseous H_2 into a small container, the pressure inside the container must be very high, like 700 $daNcm^{-2}$. In the case of liquid H_2, the pressure inside the container is usually low, around 3 $daNcm^{-2}$, but the temperature must go down to -252.77°C; and even at such low temperature, the density of H_2 is still very low, 77 kg m^{-3} [5, 30].

Crowl and Jo [31] mention the dangers and risks associated with the use of H_2. They state that danger is linked to explosive and flammable nature and risk is estimated by a combination of the accident consequences together with the probability of the accident happening.

A dangerous feature of the H_2 is its minimal ignition energy, compared to other gases, which means that H_2 is easily ignited. Furthermore, there is also the problem related to the H_2 dangerous wide range of flammability, from 4% to 77% (v/v), and, accordingly, it is important to avoid H_2 in small and unventilated spaces. Also prone to cause accidents, is the large H_2 diffusivity. Anyway, the aeration of spaces is easy due to the low molecular of H_2; for the same reason, the upward dispersion of H_2 is fast.

Materials for H_2 storage containers must be chosen based on H_2 characteristics. Operating conditions are also important to consider. Anyway, there are European Regulations that limit the choice of materials, considering safety [32]. The molecule of H_2 has a small size, which entails that the penetration of H_2 through the container walls cannot be neglected. A precaution to consider is the material of the container. If the container is constructed of metallic alloys with carbon, for example steel, then there will be likely a reaction between the carbon and the H_2, generating bubbles of CH_4, and cracks will appear on the walls. Such phenomenon is known as H_2 embrittlement, solved according to some authors [23, 33, 34] by the addition of molybdenum and chromium to the steel.

H_2 leakage is an important issue, especially if the container is at high pressure. Such leakages become dangerous because of the invisibility of the flames generated by the combustion of H_2 and the wide flammability range of H_2. Solutions to minimize these risks are the use of relief valves, rupture discs, and flame traps installed in the containers.

According to Adams *et al.* [35], the H_2 leaving by permeation a container, placed in a garage, spreads evenly, with no noticeable stratification, despite the density of H_2 being much lower than that of the air. The authors calculated the maximum flow rate by permeation for containers of common vehicles and of urban buses, considering the following parameters: (i) the air change flow rate in the garage; (ii) the age of the polymer; (iii) the temperature to which the polymer is subjected; (iv) the dimensions of the garage; and (v) the initial mass/pressure of H_2 inside the containers. The authors allowed a maximum percentage of H_2 in the garage of 1% (v/v), which, for safety reasons, is well below the lower flammability limit of H_2 in the air.

As already stated, the present chapter studies a new storage system for H_2, designed by Stenmark [31], meant to be used in vehicles. Such storage system consists of small spheres stowed randomly in a container with dimensions and shape like those of fuel tanks used in current vehicles. The purpose of this chapter is: (i) to assess the features of such storage method regarding energy stored per volume of the system (VED) and energy stored per weight of the system (GED), (ii) to calculate permeation of hydrogen from the system, (iii) to establish the best set of material for each part of the system, (iv) to check the compliance with current standards like the European Regulations [32], (v) the comparison with other storage methods for hydrogen, and (vi) to propose conceptual solutions for the filling/refilling of the system.

2.2 Outline of the New Storage System

The intended purpose of the new storage system is the feeding of propellers (internal combustion engines or stacks of PEMFC) applied in automotive propulsion. The design of this system has three main concerns, namely, (i) the maximization of the ratio of the energy content of stored H_2 to the total system weight (GED) and (ii) the maximization of the ratio of the energy content of stored H_2 to the total system volume (VED) and, (iii) to impart trust to the user.

The storage system consists of a tank, hereinafter named envelope tank, with randomly dropped spheres inside, each sphere provided with a chip. The spheres can have any diameter, as wished. The envelope tank can take any shape, for example cubic, see Figure 2.1. The spheres have H_2 inside at room temperature and at high pressure, reaching a peak of 700 daNcm^{-2}. Each sphere has a chip of a single material, for example, Si, embedded in it. This chip is a solid rectangular parallelepiped with the following dimensions (0.5 × 0.5 × 2.5 [mm]). Its function is to control the discharge/refilling of H_2 from/into the sphere. If the pressure between the spheres and the envelope tank (the pressure outside the sphere or inside the envelope tank), is less than 5 daNcm^{-2}, for example, then the chip stops the outflow of H_2 from the sphere into the envelope tank. If the working pressure is between 5 and 20 daNcm^{-2}, then the chip allows a controlled outflow of H_2 from the sphere to the envelope tank. The consumer (propeller) is fed from the envelope tank. The propeller is either a reciprocating spark-ignition engine or a stack of PEMFC.

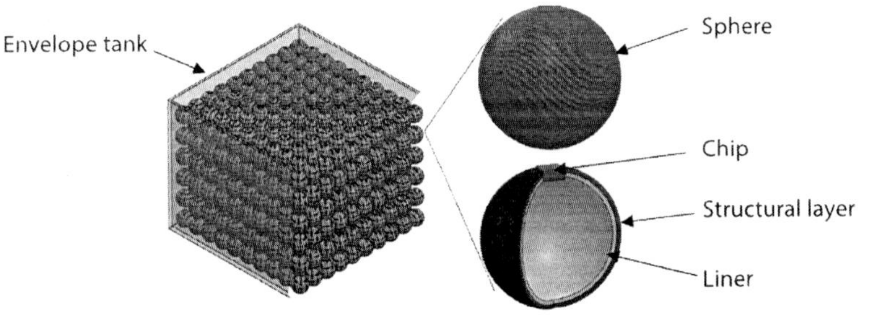

Figure 2.1 New system for hydrogen storage.

In this study, the maximum operating pressure of H_2 in the envelope tank was considered as 20 $daNcm^{-2}$.

2.2.1 Theoretical Tools Used for the System Analysis

The theoretical tools used to assess the system were:

1. Energy stored by weight (GED) and the volumetric energy density (VED);
2. The packing factor (PF);
3. Permeation of H_2 across the storage system;
4. Suitable materials for the parts of the storage system;
5. Regulations that must be complied with;
6. Filling/refilling of the spheres.

1. GED and VED

The GED, also known as gravimetric energy density, is the ratio of the energy content of hydrogen stored (in the spheres and in the spaces between these spheres) within the storage system by the total weight of the storage system (hydrogen plus spheres plus envelope tank); VED is the ratio of the energy content of hydrogen stored (in the spheres and in the spaces between these spheres) within the storage system by the outer volume of the storage system. By the way, according to Zhang *et al.* [6], for gaseous hydrogen stored at 700 $daNcm^{-2}$, which is the case for the new storage method, the target established by U.S. Department of Energy (US DoE) [36] for the GED is 1.5 $kWhkg^{-1}$ and for VED is 0.8 $kWhL^{-1}$.

To calculate the energy of the H_2 contained in the system, we have Equation (2.1),

$$E_{system} = M_{H_2} LHV_{H_2}, \tag{2.1}$$

where, LHV_{H2} is given by [5, 23], and the hydrogen mass (m_{H2}) within the storage system is given by Equation (2.2),

$$M_{H_2} = \frac{P_{in\,sph} \cdot N_{sph} \cdot V_{\text{in sph}}}{Z_{in\,sph} \cdot R_{H_2} \cdot T_{H_2}} + \frac{P_{et} . (1 - PF) V_{in\,et}}{Z_{et} \cdot R_{H_2} \cdot T_{H_2}}. \tag{2.2}$$

The compressibility values in the spheres and in the envelope tank are different because the pressures are, as well, different in these spaces.

According to the above definitions, the gravimetric energy density (GED) is, Equation (2.3),

$$\mathrm{GED} = \frac{\mathrm{E}_{\mathrm{system}}}{\left(\mathrm{M}_{\mathrm{H2}} + M_{sph} + M_{et)}\right)}, \tag{2.3}$$

and the volumetric energy density (VED) is given by Equation (2.4),

$$\mathrm{VED} = \frac{\mathrm{E}_{\mathrm{system}}}{\mathrm{V}_{\mathrm{ext\ et}}}. \tag{2.4}$$

2. *Packing Factor (PF)*

To maximize GED and VED it is necessary to maximize the number of spheres within the envelope tank. Aigueperse *et al.* [37] found that equal spheres thrown into a container are likely to occupy the space according to arrangements like crystalline structures. The simplest arrangement of spheres is named body-centered cubic (BCC) [38, 39]. According to these authors, the corresponding PF of BCC is, approximately, 0.52.

In 1611, Kepler presented a theory that originated one of David Hilbert's problems, for the three-dimensional arrangement of spheres. According to Kepler, the closed hexagonal packing has the highest PF, approximately 0.74, [38]. It was only in 1998 that Thomas Hales proposed a proof of Kepler's theory, using a software, called a proof assistant, becoming known as Isabelle HOL Light® theorem prover. This demonstration was finally accepted by the scientific community in 2017 [40].

Despite the referred assumption of Aigueperse *et al.* [37] there is no way to predict with all certainty the arrangement acquired by spheres thrown into a container. In this context, it is important to distinguish two arrangements, randomly loose packing (PRL) and randomly compact packing (PRC).

The PRC is the most compact arrangement possible for spheres thrown randomly into a container while mechanical vibration is imposed on the container [41]. Some scientific studies show that common PF values converge to 0.64 [42].

For PRL, PF is normally adopted as 0.60 [42]. However, Onoda *et al.* [43], in their study of glass spheres immersed in a liquid with negligible gravitational force, obtained a PF = 0.555.

Dong *et al.* [42], recognize that both PRC and PRL are arrangements not yet fully understood, and the existing knowledge about the subject was acquired empirically.

In Table 2.1 are given some values of PF obtained by several authors.

Pouliquen *et al.* [44] concluded that PF varies with the speed with which the spheres are introduced into the tank; the lowest PF obtained in the study was 0.62.

Donev [45] using spheres with 3.175 mm diameter within a cubic container, obtained a PF = 0.625. Using computational calculations, they

Table 2.1 Values of PF for several situations [44–48].

Reservoir	Material of spheres	Vibration	Lubricant	PF
Cylinder	Nylon	No	No	0.575
	Steel		No	0.582
Sphere	Steel		No	0.594
Cylinder	Plexiglass		No	0.605
	Steel		No	0.608
	Steel		Oil	0.611
	Steel	Yes	No	0.617
Cube	Steel		No	0.625
Cylinder	Nylon		No	0.629
Sphere	Steel		No	0.634
	Steel		No	0.635
Cylinder	Steel		Oil	0.636
	Plexiglass		No	0.636
	Steel		No	0.638
Sphere	Glass		Isopropanol	0.642
Cube	Steel		No	0.670

showed that with ellipsoids, with ratios between great and small diameters in the range of 1.2 and 1.3, it is possible to have a PF = 0.73.

Man *et al.* [46] also reached a PF of 0.642 with 11 mm glass spheres thrown into a spherical container; the container was shaken while the spheres were introduced; isopropanol was used as a lubricant to promote the accommodation of the spheres.

3. *Permeation of H_2 Across the Storage System*

The permeation flowrate of H_2 through the storage system is due to the permeability of the spheres (included the chip) and of the envelope tank. It is necessary, for safety reasons, to check if this flowrate, during a prolonged stop of the vehicle, jeopardizes the garage safety where the vehicle is parked. Therefore, it must be known the pressure, permeation flowrate of H_2, mass of H_2, and concentration of H_2 - all these over time -, for the envelope tank, the spheres, and the garage.

So, the emptying of the storage system, only by permeation, is a transient process. The calculations made, and mentioned below, are based on a chain of time intervals. The time steps are as small as required for the sake of accuracy, and during each time step it is plausible to assume a steady state.

A calculations description performed follows:

1. Mass of H_2 within a sphere is calculated by Equation (2.5),

$$\mathrm{M}_{\mathrm{H_2\text{-}sph}} = \frac{P_{in\,sph} \cdot .\mathrm{V}_{\mathrm{in\,sph}}}{Z_{in\,sph} \cdot \mathrm{R}_{\mathrm{H_2}} \cdot \mathrm{T}_{\mathrm{H_2}}}. \tag{2.5}$$

2. Concentration of H_2 in a sphere is calculated by Equation (2.6) since the mole fraction of H_2, in the spheres, is $x_{H2} = 1$.

$$[H_2]_{sph} = \frac{P_{in\,sph}}{\mathrm{Z} \cdot \mathrm{R_u} \cdot \mathrm{T}_{\mathrm{H_2}}}. \tag{2.6}$$

3. Concentration of H_2 in the envelope tank is calculated by Equation (2.7) since the mole fraction, within the envelope tank, is $x_{H2} = 1$.

$$[H_2]_{et} = \frac{P_{et}}{Z_{et} \cdot \mathrm{R_u} \cdot \mathrm{T}_{\mathrm{H_2}}}. \tag{2.7}$$

4. The calculation of the envelope tank's inner volume occupied by the spheres is given by Equation (2.8),

$$V_{et\ sph} = PF \times V_{int\ tank}. \tag{2.8}$$

5. Mass of H_2 in the volume of the envelope tank not occupied by spheres is given by (2.9),

$$M_{H_2\text{-}et} = \frac{P_{et}.(1-PF)V_{in\ et}}{Z_{et} \cdot R_{H_2} \cdot T_{H_2}} \tag{2.9}$$

6. The permeation coefficient, Φ, of H_2 was drawn from tables, Barth *et al.* [49]; its value is expressed either by mole/m^2/s/MPa$^{1/2}$, mole/m/s/MPa$^{1/2}$, or mole/m/s/MPa.

7. The H_2 solubility, S, was obtained as the quotient of the H_2 concentration by the H_2 partial pressure, if the permeation units were expressed in mole/m/s/MPa, or as the quotient of H_2 concentration by the square root of the H_2 partial pressure, if the permeation was either expressed by mole/m^2/s/MPa$^{1/2}$ or by mole/m/s/MPa$^{1/2}$. In the spheres, the H_2 partial pressure is the total pressure; likewise, the partial pressure of H_2 in the envelope tank equals the total pressure in it; the H_2 concentration of the atmosphere was plausibly assumed to be zero.

8. The diffusivity, D, of H_2 through each layer of the spheres and through the envelope tank was determined with the Equation (2.10). If the permeability is expressed in mole/m^2/s/MPa$^{1/2}$, which was the case with Si for the chip, to obtain the diffusivity in m^2/s was necessary to multiply the thickness of the layer by the permeability.

$$\Phi = DS. \tag{2.10}$$

9. Each micro-sphere is made of two concentric spheres, or layers, the smallest in the biggest, of different materials, and a parallelepipedal chip embedded in both spheres; moreover, the outer diameter of the inner sphere equals the inner diameter of the outer sphere. The inner layer, named liner, is mostly intended to provide the necessary resistance to H_2 permeation and the outer layer, named structural, is mostly intended to provide structural strength. The permeation flow of H_2 from the spheres must not be confounded with the intentional flow of H_2 from the spheres, controlled by the chip, to fuel the propeller (engine or fuel cells). The permeation occurs in two different ways: by permeation through the spheric layers and

by permeation through the chip. There is, eventually, an unwanted flow of H_2 through the interface between the chip and the sphere. This latter flow is leakage and should be as small as possible; since it depends on the quality of the manufacture of the spheres, it is human-controlled and will be neglected. Thus, the overall diffusivity for a sphere is given by (2.11): it was calculated considering the flow across the series of the composite wall of two concentric spheres (liner and structural) in parallel with the flow across the chip. As referred, $r_{o\ liner} = r_{i\ strut}$.

Thus, the calculation of the diffusivity of the sphere is given by Equation (2.11), considering the flux through the chip and the permeation flux through the composite wall of the spheres.

$$D_{Total} = \left(\frac{1}{r_{i\ liner}} - \frac{1}{r_{o\ stru}}\right)\left[\frac{D_{chip}A_{chip}}{4\pi t_{chip}} + \frac{1}{\frac{1}{D_{liner}}\left(\frac{1}{r_{i\ liner}} - \frac{1}{r_{o\ liner}}\right) + \frac{1}{D_{stru}}\left(\frac{1}{r_{i\ stru}} - \frac{1}{r_{o\ stru}}\right)}\right]. \tag{2.11}$$

10. Total resistance to the H_2 diffusion through a sphere is given by Equation (2.12).

$$R_{sph} = \frac{1}{4\pi D_{Total}}\left(\frac{1}{r_{i\ liner}} - \frac{1}{r_{o\ stru}}\right). \tag{2.12}$$

11. Mole flowrate of H_2 through a sphere is given by Equation (2.13); the H_2 concentrations are considered at the liner inner face and at the structural sphere outer face.

$$\dot{n}_{H2-sph} = \frac{[H_2]_{i\ liner} - [H_2]_{o\ stru}}{R_{sph}}. \tag{2.13}$$

12. The masses of H_2 contained in a sphere at the instant t, and at the instant t-Δt are related through the Equation (2.14).

$$M_{H2-sph}(t) = M_{H2-sph}(t-\Delta t) - \dot{n}_{H2-sph}(t) MW_{H2}\Delta t. \tag{2.14}$$

13. The H_2 concentration in the envelope tank is given by (2.7), at the inner wall, and was assumed as zero at the outer wall, exposed to the atmosphere.

14. The H_2 solubilities at the outer and inner surface of the envelope tank were calculated as indicated at point 7. The H_2 diffusivity through the envelope tank was determined by Equation (2.10).

15. It was considered a cylindrical envelope tank; Equation (2.15) gives the total resistance to the H_2 diffusion through it.

$$R_{et} = \frac{ln\left(\frac{r_{et\,o}}{r_{et\,i}}\right)}{2\pi L D_{et}}. \tag{2.15}$$

16. The H_2 mole flowrate across the envelope tank to the atmosphere is given by the Equation (2.16). The H_2 concentrations were considered at the outer and at the inner face of the envelope tank. As stated at point 13, the H_2 concentration at the outer face of the envelope tank was plausibly taken as zero.

$$\dot{n}_{H2-et} = \frac{[H_2]_{i-et} - [H_2]_{o-et}}{R_{et}}. \tag{2.16}$$

17. Mass of H_2 contained in the part of the envelope tank not occupied by spheres at the instants t and t-Δt are related with Equation (2.17).

$$M_{H2-et}(t) = M_{H2-et}(t-\Delta t) + MW_{H2}\Delta t\left[N_{sph}\dot{n}_{H2-sph}(t) - \dot{n}_{H2-et}\right]. \tag{2.17}$$

18. The perfect gas equation was used to determine the pressure in the part of the envelope tank not occupied by the spheres or inside the spheres, at a given time.

R_{H2} is the ideal gas H_2 constant, 4124 J/kg^{-1}K^{-1}; MW_{H_2} is the molecular weight of H_2, 2.016 kgkmol^{-1}; R_u is the universal constant of perfect gases; T_{H_2} is the temperature of H_2, 293.15 K; and Z is the compressibility factor of H_2 at 293.15 K and 700 daNcm^{-2}, 1.46 [50].

4. *Suitable Materials for the Parts of the Storage System*

Considering their adequacy for the storage system manufacture, a set of materials was chosen; their most important properties for the present work are in Table 2.2. The silicon (Si), mentioned in Table 2.2, was selected only for the chip.

The safety of people is always paramount. So, it was necessary to assess the risks associated with the probability of explosions and their causes. A possible cause of explosions is the accumulation of H_2 beyond a certain amount. In this work, the existing permeation was evaluated considering the materials and operating conditions. According to [56, 59], the permeability values that are available in the literature are associated with large uncertainties. Furthermore, these values are not always presented in the same units. In Table 2.2, Φ stands for permeation values at 101.3 kPa and 293.15 K. Metal alloys (Al5050-H38, SS316, Inconel 718, SS403, and W) have lower permeation values and higher strength values than polymeric materials (PP and HDPE). Composite materials (CFEP and epoxy fiberglass) have higher permeation values and greater than or equal strength values when compared to metal alloys.

Table 2.2 Properties of the materials for the sphere and chip [51–58].

Material	ρ (kgm^{-3})	σ_{yield} (MPa)	σ_{ult} (MPa)	Φ [mol/ ($msPa^{0.5}$)]	Φ [mol/ (msPa)]
Al 5050-H38	2697	220	-	4.34×10^{-20}	-
SS316	7990	290	-	1.13×10^{-18}	-
Inconel 718	8190	1100	-	1.13×10^{-17}	-
SS403	7800	310	-	4.34×10^{-20}	-
W	12750	1045	-	4.94×10^{-32}	-
Si	3220	-	-	1×10^{-8}	-
PP	870	-	17.4	-	2.6×10^{-15}
HDPE	1275	-	27	-	8.98×10^{-16}
CFEP	1790	-	4000	-	1.9×10^{-16}

Therefore, the values of the Table 2.2 suggest the idea of coupling two materials in the sphere construction: two concentric spherical layers, the outer layer to give strength and the inner layer to give resistance to H_2 permeation.

The materials for all the parts of the storage system were combined according to Table 2.3, from the set of materials of Table 2.2. For all combinations, CFEP was used for the structural layer, and Si was used for the chip. As shown below, each one of these combinations was later tested regarding its performance on H_2 permeation.

The Si was chosen for the chip owing to its easy tooling [60]; because of this, it is a commonly used material for the manufacture of electronic devices.

Following Dwivedi *et al.* [61], materials with high strength (titanium, steel with a high magnesium content, magnesium alloys, and steel) are susceptible to embrittlement caused by H_2, if subjected to high temperature or high pressure.

Table 2.3 Combinations of materials for the storage system (CFEP for all structural layers) [63].

<table>
<tr><th>Combinations</th><th>Envelope tank</th><th>Lining</th></tr>
<tr><td>1</td><td>Al 5050-H38</td><td rowspan="6">Al 5050-H38</td></tr>
<tr><td>2</td><td>SS316</td></tr>
<tr><td>3</td><td>SS403</td></tr>
<tr><td>4</td><td>Inconel 718</td></tr>
<tr><td>5</td><td>PP</td></tr>
<tr><td>6</td><td>HDPE</td></tr>
<tr><td>7</td><td>Al 5050-H38</td><td rowspan="6">SS316</td></tr>
<tr><td>8</td><td>SS316</td></tr>
<tr><td>9</td><td>SS403</td></tr>
<tr><td>10</td><td>Inconel 718</td></tr>
<tr><td>11</td><td>PP</td></tr>
<tr><td>12</td><td>HDPE</td></tr>
</table>

(Continued)

Table 2.3 Combinations of materials for the storage system (CFEP for all structural layers) [63]. (*Continued*)

Combinations	Envelope tank	Lining
13	Al 5050-H38	SS403
14	SS316	
15	SS403	
16	Inconel 718	
17	PP	
18	HDPE	
19	Al 5050-H38	W
20	SS316	
21	SS403	
22	Inconel 718	
23	PP	
24	HDPE	
25	Al 5050-H38	PP
26	SS316	
27	SS403	
28	Inconel 718	
29	PP	
30	HDPE	

The thickness of the structural layer was determined using the theory of thin-walled reservoirs, and a safety factor equal to 1.41, according to [62].

5. *Regulations to be Complied With*

The European Regulations on the approval of future H_2-powered vehicles is the EU 406/2010 [32]. The standard allows the neglecting of permeation on metallic container materials. However, polymeric materials must undergo tests to check their permeation performance.

At steady state, the highest allowable permeation flow rate is 6 $Ncm^3h^{-1}L^{-1}$ (per liter of the internal volume of the envelope tank). For safety reasons, in closed spaces, like a garage, it is necessary to guarantee a minimum air renewal per hour of 0.03 [64] to avoid flammable mixtures between H_2 and air.

The value of $C_{\%}$ must be lower than 1%, Equation (2.18), where $C_{\%}$ represents the ratio of the volumetric outflow of the H_2 leaving, by permeation, the envelope tank, Q_{H2}, and the sum of the inflow of air in the space due to air renewal, Q_{air} with Q_{H2}. So,

$$C_{\%} = \frac{100 \cdot Q_{H2}}{Q_{air} + Q_{H2}}. \tag{2.18}$$

In the current study were considered the three following scenarios proposed by Adams *et al.* [35] for domestic garages with parked vehicles inside; see Table 2.4.

6. *Filling/Refilling of Spheres*

A successful dissemination of this storage system depends on the availability of an appropriate supply network for consumers: owners of tourist or heavy-duty vehicles should easily refuel their vehicles when needed. Attached with this topic there are lots of questions to be addressed, among which stands out how and where to fill/refill the spheres.

This paper essays two reasonable conceptual solutions for such question, see Figures 2.2 and 2.3, which presupposes that envelope tanks are removable and interchangeable, so that when spheres are discharged or contain H_2 at too low pressure, envelope tanks can be removed from the vehicle where they are and then replaced by others, with the same dimensions, but

Table 2.4 Scenarios for domestic garages with parked vehicles, adapted from Adams *et al.* [35].

Scenarios	Features				
	Garage volume (m^3)	Garage free volume (m^3)	Volume of impermeable material (m^3)	Natural ventilation of the garage (ACH)	Natural ventilation of the garage (m^3/h)
1	50	46	4	0.03	1.38
2	33	31	2	0.03	0.93
3	19	18	1	0.03	0.54

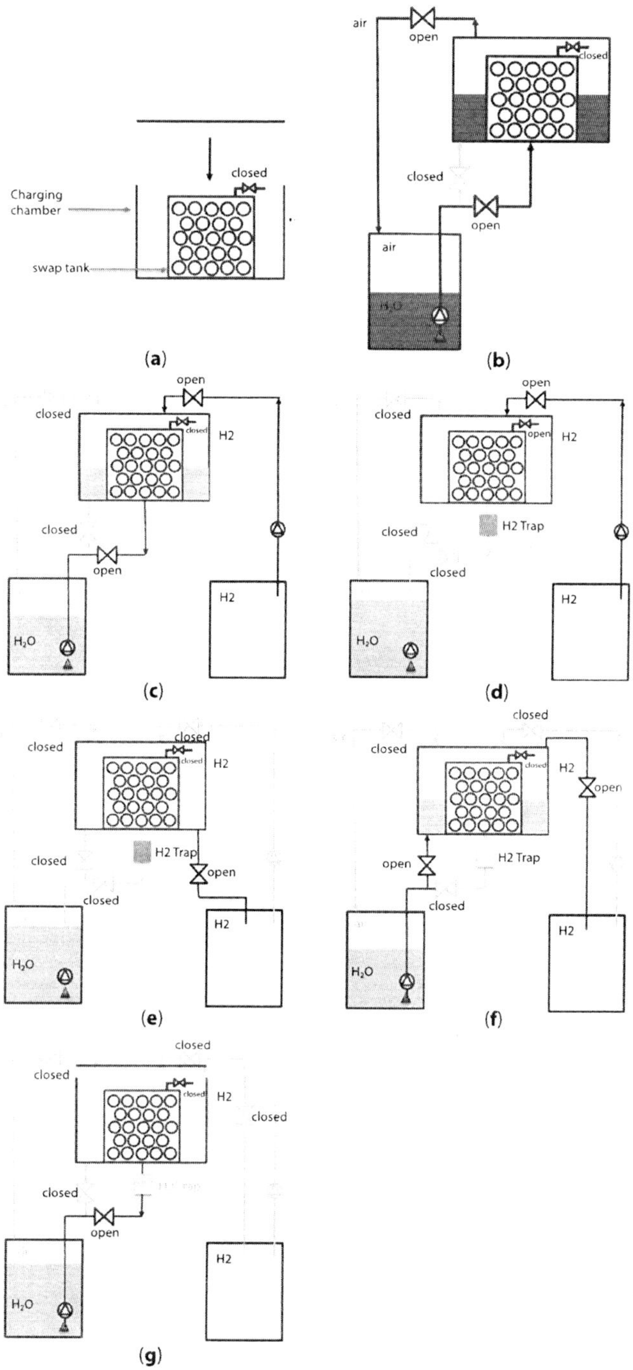

Figure 2.2 Charging chamber process – first alternative.

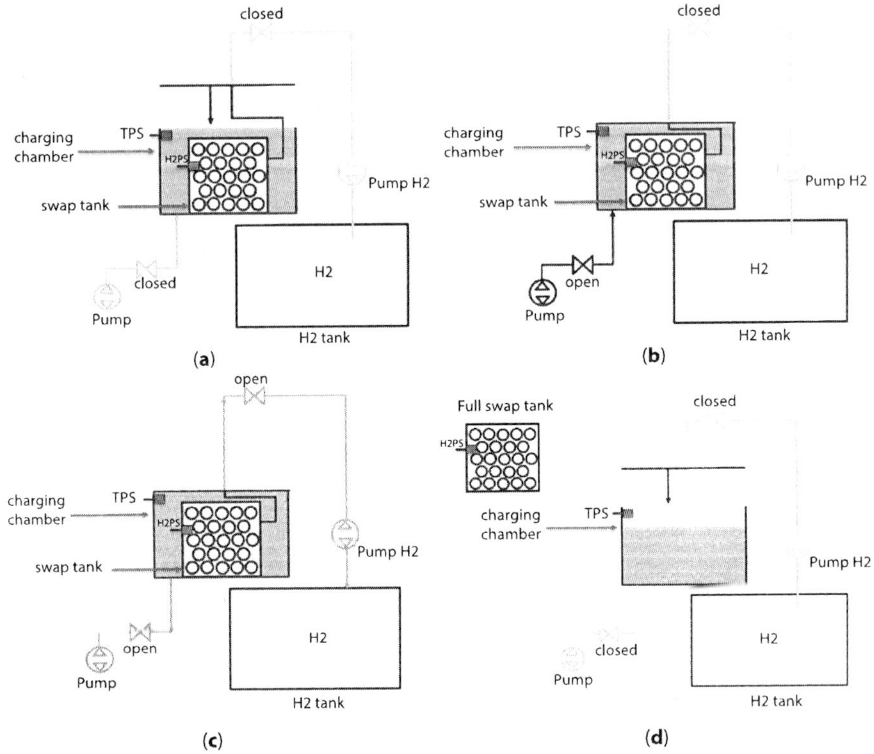

Figure 2.3 Charging chamber process – second alternative, TPS; Tank pressure switch; H2PS; H2 pressure switch; —— Connection; —— Inactive; —— H_2 injection; —— H_2 removal; —— air injection; —— air removal.

whose spheres are already filled with H_2. This replacement can be done at service stations, and the customer will pay according to the difference of the amount of H_2 between the new (charged) and the replaced (discharged) envelope tank; the amount of H_2 can be determined by gross weight minus tank tare or through the H_2 pressure. It should be noted that one of the major concerns on the basis of the conception of both solutions was to keep the H_2 totally unmixed with air, as it is known that the flammability range of H_2 is very wide.

The filling/refilling of the storage system can be done according to the two alternative procedures described below. As the first alternative:

1. The envelope tank is placed in a reservoir, the charging chamber, dug in the ground Figure 2.2 (a), by means of a

crane or gantry, for example. The shut-off valve attached to the envelope tank is closed.

2. A lid closes the charging chamber. Water is introduced in the charging chamber and expels the air Figure 2.2 (b). The shut-off valve attached to the envelope tank remains closed.
3. Right after the charging chamber is full of water, H_2 is introduced, from an underground reservoir (or a cavern as proposed by Tarkowski [29]) into the charging chamber, expelling the water Figure 2.2 (c).
4. Without water in the charging chamber, the H_2 will be compressed into de charging chamber up to a pre-set high pressure; the chip allows the filling of the sphere on which it is embedded; shut-off valve attached to the envelope tank is open to equalize the pressure within and without the envelope tank; a H_2 trap permits only the flow of water and prevents the outflow of H_2 from the charging chamber Figure 2.2 (d).
5. When the pre-set pressure is reached inside the spheres, the charging chamber and the envelope tank are both depressurized (shut-off valve of envelope tank still open), down to the maximum pressure desired in the envelope tank, say 20 bar, Figure 2.2 (e).
6. Shut-off valve of envelope tank is closed. Water is pumped into the charging chamber to expel the remaining H_2 from there to the underground reservoir (or cavern) of H_2 Figure 2.2 (f).
7. The lid of the charging chamber is lifted; the water in the charging chamber falls back to its reservoir Figure 2.2 (g).

As the second alternative, the filling/refilling of the storage system can be done as described below:

1. The envelope tank is placed in the charging chamber, dug in the ground, by means of a crane or gantry, for example. Then, a lid closes the charging chamber. There is a hose that links the lid to the envelope tank, Figure 2.3 (a).
2. Water is compressed into the charging chamber, while H_2 is compressed into the envelope tank. To prevent the collapse of the envelope tank, the pressure of H_2 in the envelope tank must equal the pressure of water in the charging

chamber. The chip allows the filling of the sphere on which it is embedded Figure 2.3 (b).
3. When the desired pressure is reached inside the spheres, the charging chamber and the envelope tank are both depressurized, the envelope tank down to the maximum pressure desired in the envelope tank, say 20 bar, into the H_2 reservoir (or cavern) and the charging chamber down to the atmospheric pressure Figure 2.3 (c).
4. The lid of the charging chamber is removed Figure 2.3 (d).

To compare the costs of the two filling alternatives mentioned, it is only relevant to consider the stage of pressure build-up. Indeed, at alternative 1, the energy consumption is: at step 1 – energy of the crane or gantry; at step 2 – energy of the water pump, that depends on the head loss and the water flowrate on the pump; at step 3 – negligible, since the underground reservoir must be at a medium pressure, high enough to warrant the filling of the charging chamber up to a low pressure, say 5 bar; at step 4 – significant energy consumption by the compressor of H_2; at step 5 – negligible energy consumption on valve commands, since there is just pressure relief; at step 6 – like step 2; at step 7 – like step 1. For the filling alternative 2, the energy consumption is: at step 1 – energy of the crane or gantry; at step 2 – significant energy consumption by the compression of water and by the compressor of H_2; at step 3 – negligible energy consumption on valve commands, since there is just pressure relief; at step 4 – like step 1. So, the energy consumption of alternative 1 at step 4 was compared with the energy consumption of alternative 2 at step 2. Energies consumed by cranes or gantries were considered similar for both alternatives.

For the energy consumption to fill/refill an envelope tank by the first alternative, it was assumed: specific heats constant for H_2; a continuous charging flowrate either of H_2; no heat losses from the envelope tank or the charging chamber; thermodynamic equilibrium during the pressure build-up. According the first law of Thermodynamics, in the process of filling a reservoir with a pressurized gas, the amount of energy supplied by the gas compressor equals the increase in internal energy of the gas accumulated in the reservoir. So,

$$E_{fill/refill} = \quad U = [M_{cc+et}(0) + m_{H2}t]c_v T - M_{cc+et}(0)c_v T(0). \quad (2.19)$$

In Equation (2.19), $M_{cc\text{-}et}(0)$ is the mass of H_2 in the charging chamber and the envelope tank just at the beginning of step 4 of the procedure; m_{H2} is the mass flowrate of H_2 during the same step 4 of the procedure.

Besides, from this equation can be drawn that as a gas is pressurized into a reservoir, the ratio of the absolute temperature of the gas within the reservoir by the absolute temperature of the gas at the inlet of the reservoir approaches c_p/c_v of the gas as the time goes by. For H_2 (c_p = 14.31 kJ/kg/K and c_v = 10.19 kJ/kg/K at 300 K), the temperature of the H_2 tends to 420 K.

In the case of the second alternative, the same Equation (2.19) was used for the filling/refilling of the envelope tank; the energy consumption to pressurize the water contained in the space between the envelope tank and the charging chamber was calculated with

$$E_{water\ fill/refill} = \Delta p V_{water\ pumped}. \tag{2.20}$$

The pressurization of the water starts as the space between the changing chamber and the envelope tank is already filled with water at atmospheric pressure; the mass of water to be pumped in order to increase the pressure of water from the atmospheric pressure up to the desired pressure is $(V_{cc} - V_{et}).(\rho_{final} - \rho_{initial})$.

In both alternatives, the volume of the charging chamber was assumed to be 10% higher than the volume of the envelope tank.

2.3 Results

For all the combination of materials listed in Table 2.3, the H_2 mass in each sphere and the envelope tank were calculated over time, with Equations (2.5) and (2.9), respectively.

The time-step of 5000 s was used in all the calculations. During each time-step, both pressures P_{et} and P_{sph} were considered constant.

The time-step was set considering two conflicting notions: large time-steps avoid fastidious and time-consuming calculations and small time-steps avoid inaccurate results. The value of 5000 s, as the time-step, results from a trade-off between those two conflicting notions.

The pressure considered for calculating the permeation of the two layers of spheres was taken as the difference between the pressures within and without the spheres; similarly, the pressure considered for calculating the permeation of the envelope tank was taken as the difference between the pressures within the envelope tank and the atmospheric pressure.

The atmospheric temperature and the H_2 temperature in the envelope tank and inside the spheres were set at 293.15 K. The atmospheric pressure was set at 101325 Pa. The initial pressure inside the spheres was set at 71 MPa; the initial pressure in space between spheres of the envelope tank was set at 0.5 MPa. A PF of 0.52 was considered since this value is usually found in the randomly stackings of spheres.

Based on the combinations in Table 2.3, calculations were performed to find the best combination corresponding to the lowest permeation loss of the spheres to the garage. To do this, it was considered a situation with a car parked inside a garage, being the fuel-storage system of the car filled with H_2 at 700 daNcm^{-2} within the spheres and at 5 bar in the space of the envelope tank between spheres. Under these conditions, H_2 will naturally permeate from the spheres to the tank and thence to the garage. The calculation aimed to obtain the time required for reaching 20 bar in the envelope tank between spheres, desirably the longest possible, and at the same time to check if the value of $C_{\%}$, at that moment, was below 1%, as required by regulations.

To obtain results comparable to those of Adams *et al.* [35], it was assumed the storage system mounted (i) in large cars, parked in garages with a volume of 50 m^3, a free volume of 46 m^3 and 0.03 ACH (1.38 m^3/h); (ii) in small cars, parked in garages with a volume of 33 m^3, a free volume of 31 m^3 and 0.03 ACH (0.93 m^3/h); and (iii) in micro cars, parked in garages with a volume of 19 m^3, a free volume of 46 m^3 and 0.03 ACH (0.54 m^3/h) – see Table 2.4.

The results, see Table 2.5, show that in all three scenarios of Table 2.4, the value of $C_{\%}$ is well below 1%, as wished. It also turns out that it takes about 2.5 months for the pressure in the envelope tank to rise from 5 to 20 bar, except for combinations 25 to 30, owing to the material of the lining being PP, whose permeation resistance is very low.

Further analysis was made for combinations 1, 7, and 19 since these combinations correspond to the lowest values of $C_{\%}$ and the longer elapsed time – see Figure 2.4 and Table 2.5. As an instance, for combination 1, the progress of H_2 pressure over time, either in spheres or in the envelope tank, is shown in Figure 2.4; combinations 7 and 19 have graphs quite like Figure 2.4. For the same combination 1, the progress of $C_{\%}$ over time is shown in Figure 2.5. It must be stressed that calculations were performed when the car is parked and the propeller off, as the pressure in the envelope tank rises from 5 up to 20 bar.

Graph of Figure 2.4 shows that as the pressure inside the spheres decreases the pressure in the envelope tank increases, despite the envelope tank permeation to the garage. The explanation lies in the facts (i) that the

Table 2.5 Time to get 20 bar in the envelope tank and values of $C_{\%}$ for the scenarios of Table 2.4.

Combinations	Elapsed time (days)	$C_{\%} = 100 \times Q_{H2}/(Q_{air}+Q_{H2})$ in garage		
		Scenarios		
		1	2	3
1	79.6	$2.8 \cdot 10^{-10}$	$4.1 \cdot 10^{-10}$	$7.1 \cdot 10^{-10}$
2	79.6	$1.1 \cdot 10^{-8}$	$1.6 \cdot 10^{-8}$	$2.7 \cdot 10^{-8}$
3	79.6	$1.4 \cdot 10^{-5}$	$2.1 \cdot 10^{-5}$	$3.6 \cdot 10^{-5}$
4	79.6	$3.8 \cdot 10^{-7}$	$5.6 \cdot 10^{-7}$	$9.7 \cdot 10^{-7}$
5	82.6	$3.1 \cdot 10^{-3}$	$4.7 \cdot 10^{-3}$	$8.0 \cdot 10^{-3}$
6	80.4	$8.3 \cdot 10^{-4}$	$1.2 \cdot 10^{-3}$	$2.1 \cdot 10^{-3}$
7	79.6	$2.8 \cdot 10^{-10}$	$4.1 \cdot 10^{-10}$	$7.1 \cdot 10^{-10}$
8	79.6	$1.1 \cdot 10^{-8}$	$1.6 \cdot 10^{-8}$	$2.7 \cdot 10^{-8}$
9	79.6	$1.4 \cdot 10^{-5}$	$2.1 \cdot 10^{-5}$	$3.6 \cdot 10^{-5}$
10	79.6	$3.8 \cdot 10^{-7}$	$5.6 \cdot 10^{-7}$	$9.7 \cdot 10^{-7}$
11	82.6	$3.1 \cdot 10^{-3}$	$4.7 \cdot 10^{-3}$	$8.0 \cdot 10^{-3}$
12	80.4	$8.3 \cdot 10^{-4}$	$1.2 \cdot 10^{-3}$	$2.1 \cdot 10^{-3}$
13	69.4	$2.8 \cdot 10^{-10}$	$4.1 \cdot 10^{-10}$	$7.1 \cdot 10^{-10}$
14	69.4	$1.1 \cdot 10^{-8}$	$1.6 \cdot 10^{-8}$	$2.7 \cdot 10^{-8}$
15	69,4	$1.4 \cdot 10^{-5}$	$2.1 \cdot 10^{-5}$	$3.6 \cdot 10^{-5}$
16	69,4	$3.8 \cdot 10^{-7}$	$5.6 \cdot 10^{-7}$	$9.7 \cdot 10^{-7}$
17	71,8	$3.1 \cdot 10^{-3}$	$4.7 \cdot 10^{-3}$	$8.0 \cdot 10^{-3}$
18	70,0	$8.3 \cdot 10^{-4}$	$1.2 \cdot 10^{-3}$	$2.1 \cdot 10^{-3}$
19	79.6	$2.8 \cdot 10^{-10}$	$4.1 \cdot 10^{-10}$	$7.1 \cdot 10^{-10}$
20	79.6	$1.1 \cdot 10^{-8}$	$1.6 \cdot 10^{-8}$	$2.7 \cdot 10^{-8}$
21	79.6	$1.4 \cdot 10^{-5}$	$2.1 \cdot 10^{-5}$	$3.6 \cdot 10^{-5}$
22	79.6	$3.8 \cdot 10^{-7}$	$5.6 \cdot 10^{-7}$	$9.7 \cdot 10^{-7}$

(Continued)

Table 2.5 Time to get 20 bar in the envelope tank and values of $C_{\%}$ for the scenarios of Table 2.4. (*Continued*)

Combinations	Elapsed time (days)	$C_{\%} = 100\times Q_{H2}/(Q_{air}+Q_{H2})$ in garage		
		Scenarios		
		1	2	3
23	82.6	$3.1\cdot10^{-3}$	$4.7\cdot10^{-3}$	$8.0\cdot10^{-3}$
24	80.4	$8.3\cdot10^{-4}$	$1.2\cdot10^{-3}$	$2.1\cdot10^{-3}$
25	0.8	$2.7\cdot10^{-10}$	$4.0\cdot10^{-10}$	$6.8\cdot10^{-10}$
26	0.8	$1.0\cdot10^{-8}$	$1.5\cdot10^{-8}$	$2.6\cdot10^{-8}$
27	0.8	$1.4\cdot10^{-5}$	$2.0\cdot10^{-5}$	$3.5\cdot10^{-5}$
28	0.8	$3.7\cdot10^{-7}$	$5.4\cdot10^{-7}$	$9.4\cdot10^{-7}$
29	0.8	$3.0\cdot10^{-3}$	$4.5\cdot10^{-3}$	$7.8\cdot10^{-3}$
30	0.8	$8.1\cdot10^{-4}$	$1.2\cdot10^{-3}$	$2.1\cdot10^{-3}$

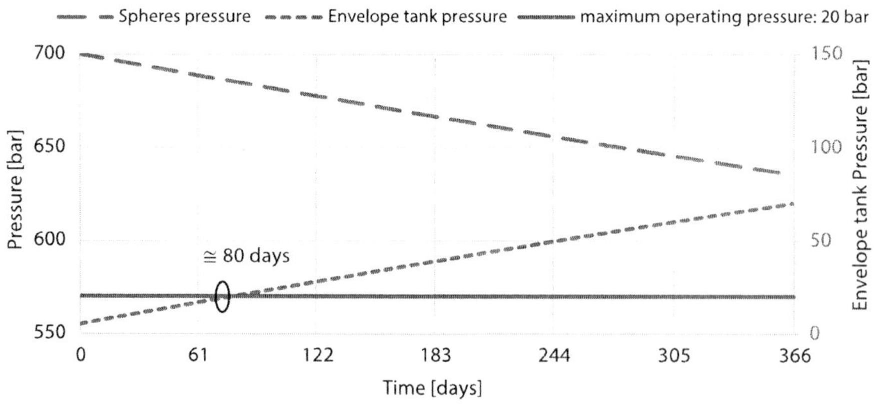

Figure 2.4 Pressure inside the spheres and in the envelope tank, combination 1 from Table 2.5.

envelope tank permeation values are lower than the spheres permeation values and (ii) the pressure in the spheres is far above the pressure in the envelope tank.

Graph of Figure 2.5 proves that for all the three scenarios of Table 2.4, the new storage system guarantees maximum values of $C_{\%}$, even for

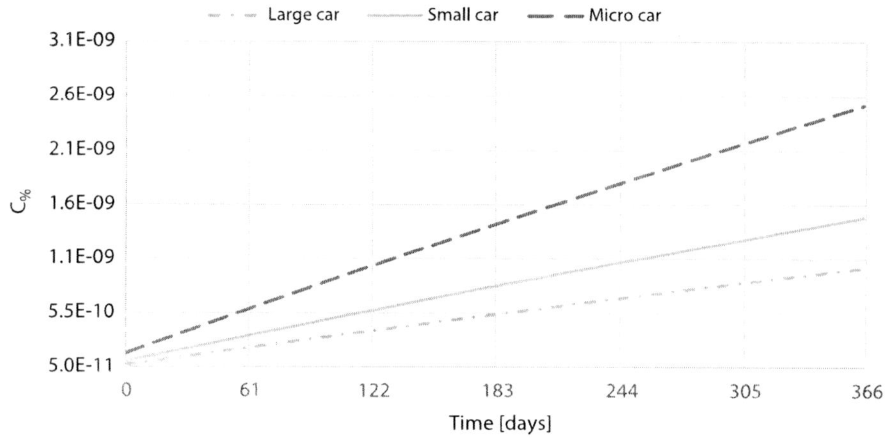

Figure 2.5 $C_{\%}$ in the garage, combination 1 from Table 2.5.

1 year of parking, far below the 1% required by European Regulations [32]. Thus, it seems that this storage system is suitable for the everyday use of hydrogen-propelled cars.

Figure 2.6 shows GED obtained through Equation (2.3), for a pressures of 700 daNcm^{-2} inside the spheres, and 20 daNcm^{-2} inside the envelope

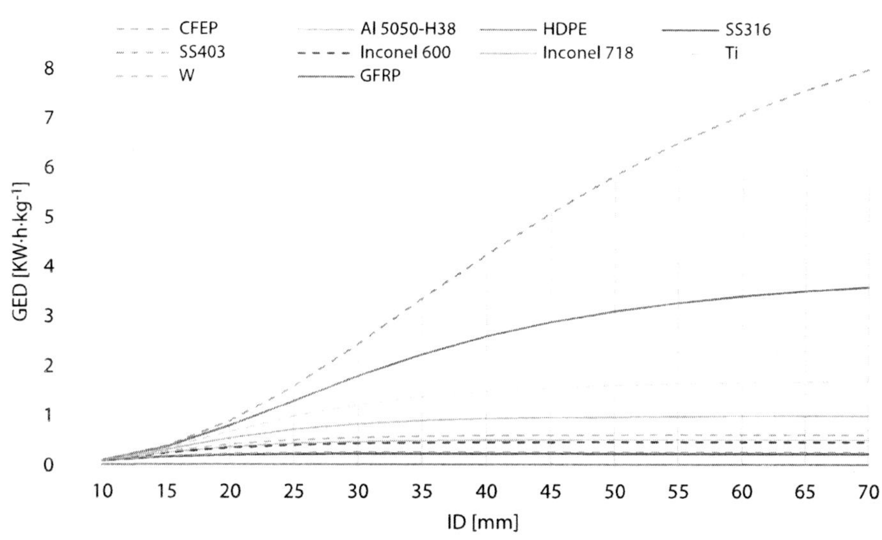

Figure 2.6 GED of spheres made of several materials with ID from 10 up to 70 mm, at 700 daNcm^{-2} of inside pressure.

tank; several materials for the spheres were assumed; the inner diameters of the spheres (ID) were considered in the range of 10 up to 70 mm.

For spheres with a determined ID and a determined pressure inside, the energy of the H_2 contained in the storage system is obtained by Equation (2.1). The GED increases with the strength of the materials used in the manufacture of the spheres (the weight of the spheres lowers). Besides, the GED increases with the ID because GED is proportional to the cube of ID. From Figure 2.6, composite materials (CFEP and GFRP) have values of GED 3 up to 8 times larger than polymers or metals.

Figure 2.7 shows values of VED obtained through Equation (2.4). For a given pressure and given ID, the VED increases as spheres are manufactured with more resistant materials (volume of the spheres drops). Figure 2.7 shows that composite materials (CFEP and GFRP) have the higher values of VED (lower thicknesses of spheres).

The Table 2.6 presents the most important features of the new storage system (envelope tank and spheres comprised), being the ID of spheres 70 mm; the packing factor was either assumed 0.52, 0.60, 0.63 or 0.74; the safety factor (CS) was taken as 1.41; the inner volume of the envelope tank was considered as 0.122 m^3; two types of chip material were considered. For the calculations in Table 2.6, aluminum was considered for the liner material, CFEP for the structural layer and aluminum for the envelope tank.

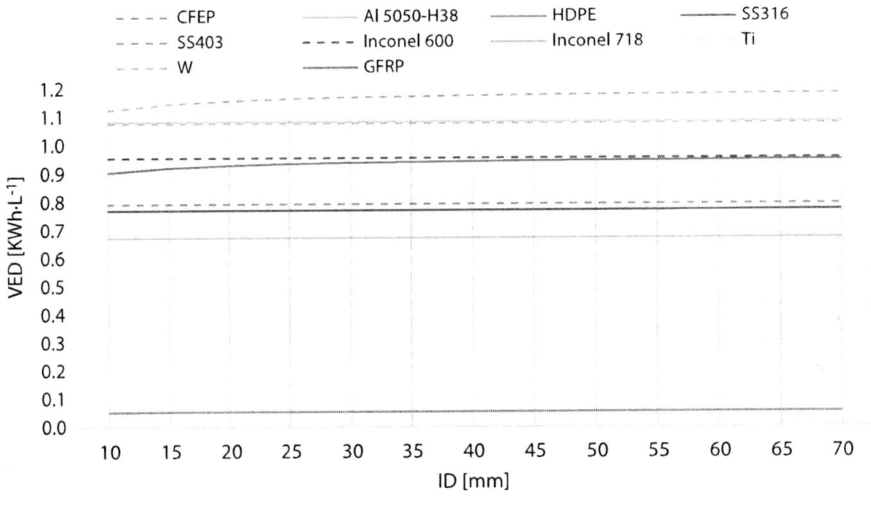

Figure 2.7 VED of spheres made of several materials with ID from 10 up to 70 mm, at 700 daNcm^{-2} of inside pressure.

Table 2.6 Characteristics of the storage system (envelope tank + spheres + chip), for different packing factors, CS 1.41.

Chip material	**SS316**				**Si**			
Packing factor	**0.52**	**0.60**	**0.63**	**0.74**	**0.52**	**0.60**	**0.63**	**0.74**
Number of spheres	338	390	410	482	338	390	410	482
GED_{syst} [$kWhkg^{-1}$]	6.16	6.33	6.38	6.55	6.81	7.02	7.09	7.30
VED_{syst} [$kWhL^{-1}$]	0.63	0.72	0.75	0.88	0.63	0.72	0.75	0.88
Mass of the system [kg]	12.54	13.98	14.53	16.52	11.34	12.60	13.08	14.82
Total mass of H_2 (spheres + envelop tank) [kg]	2.31	2.66	2.78	3.25	2.31	2.66	2.78	3.25
Useful mass of H_2 (spheres) [kg]	2.25	2.60	2.73	3.21	2.25	2.60	2.73	3.21

In Figure 2.8 (a) and (b) are compared the new storage system (named Sfeers) and two actual and common passenger cars: Toyota Mirai and Hyundai Nexo. The GED of the new storage system almost trebles the GED for Toyota Mirai and Hyundai. On the other hand, as can be seen in Figure 2.8 (c) and (d), the VED and the total mass of H_2 stored is approximately half of the counterpart values for Toyota Mirai and Hyundai Nexo; the VED is a major disadvantage of the new storage system against the actual hydrogen propelled passenger vehicles. In the case of Mirai and Nexo, the values shown are provided by the carmakers, whilst the values of the new storage system (Sfeers) have range bars due to the various values adopted to the packing factor and the density of the chip material [65, 66].

Regarding energy consumption for the filling/refilling, the data at Table 2.7 were assumed for both refilling alternatives. It was taken an inner diameter of 40 mm and 2 mm of thickness for the spheres, roughly halfway of the diameters considered in the preceding calculations. The filling/refilling

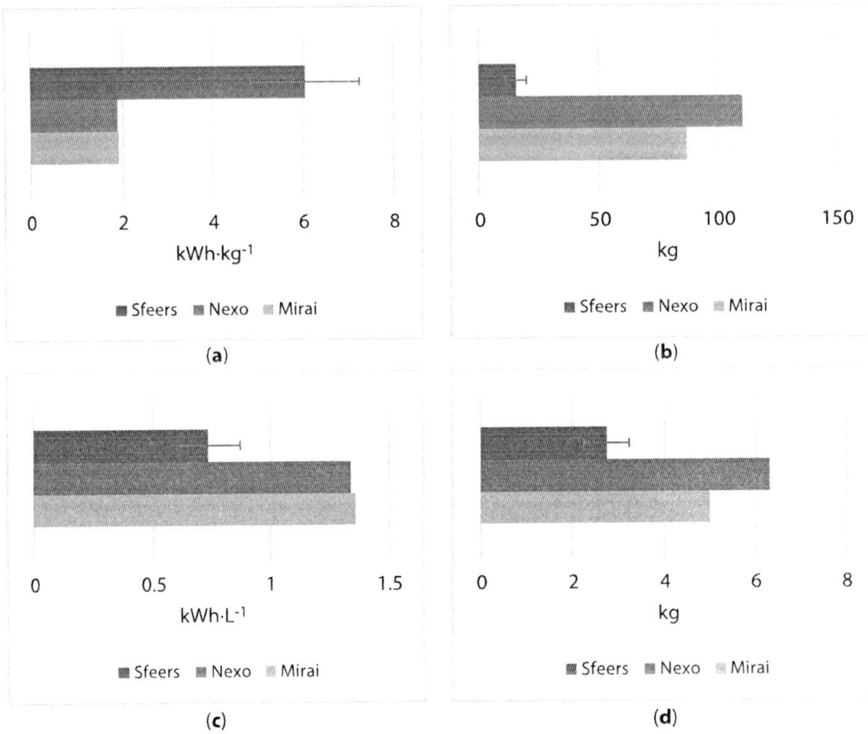

Figure 2.8 Comparison between the system under study and the storage systems of Toyota Mirai and Hyundai Nexo, regarding GED (a), total system mass (b), VED (c) and mass of stored hydrogen (d).

time of 3600 s, and the 10% of excess volume of the charging chamber in relation to the outer volume of the envelope tank were arbitrarily taken. To make calculations easier, a compressibility factor Z = 1.4 for H_2 in all situations.

The following results, see Table 2.8 were obtained for both alternatives. Alternative 2 for the filling/refilling of the envelope tank is, regarding energy consumption, slightly cheaper than alternative 1, because it is easier to compress a liquid (water) than a gas (H_2), in the charging chamber around the envelope tank.

2.4 Conclusions

The storage system of H_2, treated in this chapter, is safer than the usual storage of H_2 in containers at high pressure. This is because the mass of

Table 2.7 Data for the evaluation of energy consumption during refilling.

Data	Units	
Initial pressure	MPa	2
Final pressure	MPa	70
Temperature	K	300
Volume envelope tank	m^3	1
Inner diameter of spheres	mm	40
Thickness of spheres	mm	2
Volume of charging chamber		1.1
Time for refilling	s	3600
LHV of H_2	MJ/kg	119.96

Table 2.8 Energy consumption during refilling.

Results	Units	Alternative 1	Alternative 2
Initial mass of H_2 in the envelope tank	kg	1.02	1.02
Final mass of H_2 in the envelope tank	kg	16.55	16.55
Energy consumed by refilling/ Energy contained	%	8.6	7.7

hydrogen is divided into several and small spheres, containing each one harmless amounts of hydrogen in case of the bursting of the spheres; obviously it is unlikely that all spheres within the envelope tank will burst simultaneously. The best feature of this system, when compared with hydrogen conventional gas pressurized systems, is the GED, whereas it performs mediocrely concerning the VED. Anyway, it is above the target set by US DoE of GED = 1.5 $kWhkg^{-1}$ and of VED = 0.8 $kWhL^{-1}$.

Since both GED and VED increase with the ID of spheres, the choice for a determined ID must be substantiated on, for example, the maximum amount of hydrogen, released by the bursting of a sphere, which we are willing to accept from the point of view of our safety.

The packing factor must be as high as possible to increase GED and VED. This is achievable through vibration of the envelope tank, lubrication between spheres, the best speed at which the spheres are thrown into the envelope tank, and the mixture of spheres with different diameters.

The best set of materials for the storage system (chip + sphere + envelope tank) was studied having in mind the permeation flowrate. All the combinations of materials, besides the PP, permit the safe use of the storage system. After careful analysis, it is verified that the best option, amid the materials selected, consists of aluminum for the liner of the spheres and the envelope tank, CFEP for the structural layer of the spheres, and Si for the chip. Moreover, if European Regulations are accepted, and following the procedure of Adams *et al.* [35], if the car is parked in a garage for about 2.5 months, the storage system will never create a dangerous situation concerning the safety of persons or goods.

Analytical [67, 68] and experimental [68, 69] studies about permeation flowrate from CGH_2 vessels and about the concentration of H_2 around such vessels, as in garages, gave non-discrepant results with those of the present study.

Obviously, some of the results now presented can be slightly corrected in the future, as the data available on the permeation of materials is still a topic that needs further study. There is no large, well-consolidated base about this information. Yet, it is not likely that such corrections might call into question the present ones.

At last, regarding the preferable process for the envelope tank refilling, the second alternative is conceptually the simplest and its energy consumption slightly lower than the energy consumption of the other alternative.

Abbreviations

List of Symbols

A	Area
ACH	Air changes per hour
C	Ratio of flowrate of H_2 and flowrate of H_2 and air
CGH_2	Compressed gas hydrogen
CS	Safety factor
D	Diffusivity

E	Energy
GED	Gravimetric energy density
L	Length
LHV	Low heating value
M	Mass
MW	Molecular weight
n	Permeation mole flowrate
N	Number of
P	Pressure
PEMFC	Proton exchange membrane fuel cell
PF	Packing Factor
Q	Flowrate
r	Radius
R	Gas constant, resistance to diffusion
S	Solubility
T	Temperature
t	Time
U	Internal energy
V	Volume
VED	Volumetric energy density
x	Mole fraction
Z	Compressibility factor

Subscripts

cc	Charging chamber
et	Envelope tank
et sph	Spheres of the envelope tank
ext et	Exterior of envelope tank
i	Inner
int all sph	Inside all spheres
in et	Inside envelope tank
in sph	Inside the sphere
int tank	Inside the tank
ext tank	Outer tank
o	Outer
p-H_2	Maximum H_2 allowable permeation
sph	Sphere
strut	Structural layer

ult	Ultimate
u	Universal
yield	Yield
%	Percentage
air	Air
et	Envelope tank
H_2	Hydrogen

Greek Symbols

Δ	Variation
ρ	Density
σ	Stress
Φ	Permeation coefficient

References

1. World Bank, Fossil fuel energy consumption. Data, Energy Foss Fuels Consum, 2016, Available online: http://data.worldbank.org/indicator/EG.USE.COMM.FO.ZS (accessed on 20 September 2021).
2. Strauss, M., EU to boost green hydrogen use for decarbonisation, focus on energy efficiency, Reuters, Available online: https://www.reuters.com/article/us-climate-change-eu-hydrogen/eu-to-boost-green-hydrogen-use-for-decarbonisationfocus-on-energy-efficiency-idUSKBN2491JA (accessed on 15 February 2021).
3. European Council of the European Union, Available online: https://www.consilium.europa.eu/en/press/press-releases/2020/12/18/paris-agreement-council-transmits-ndc-submission-on-behalf-of-eu-and-member-states/ (accessed on 11 February 2021).
4. Herrero, S.T., Nicholls, L., Strengers, Y., Smart home technologies in everyday life: Do they address key energy challenges in households? *Curr. Opin. Environ. Sustain.*, 31, 65–70, 2018.
5. Spiegel, C., *Designing and building fuel cells*, 1st ed., McGraw-Hill, New York, USA, 2007.
6. Zhang, F., Zhao, P., Niu, M., Maddy, J., The survey of key technologies in hydrogen energy storage. *Elsevier, Int. J. Hydrogen Energy*, 41, 14535–14552, 2016.
7. Acar, C., Bicer, Y., Demir, M.E., Dincer, I., Transition to a new era with light-based hydrogen production for a carbon-free society: An overview. *Int. J. Hydrogen Energy*, 44, 47, 25347–25364, 2019.

8. Ren, J., Gao, S., Tan, S., Dong, L., Hydrogen economy in China: Strengths–weaknesses–opportunities–threats analysis and strategies prioritization. *Renewable Sustainable Energy Rev.*, 41, 1230–1243, 2015.
9. Ren, J., Gao, S., Tan, S., Dong, L., Scipioni, A., Mazzi, A., Role prioritization of hydrogen production technologies for promoting hydrogen economy in the current state of China. *Renewable Sustainable Energy Rev.*, 41, 1217–1229, 2015.
10. Lu, J., Zahedi, A., Yang, C., Wang, M., Peng, B., Building the hydrogen economy in China: Drivers, resources and technologies. *Renewable Sustain. Energy*, 23, 543–556, 2013.
11. Ball, M., Wietschel, M., Rentz, O., Integration of a hydrogen economy into the German energy system: An optimising modelling approach. *Int. J. Hydrogen Energy*, 32, 1355–1368, 2007.
12. Lee, D.-H., Development and environmental impact of hydrogen supply chain in Japan: Assessment by the CGELCA method in Japan with a discussion of the importance of the importance of biohydrogen. *Int. J. Hydrogen Energy*, 39, 19294–19310, 2014.
13. Lattin, W. and Utgikar, V., Transition to hydrogen economy in the United States: A 2006 status report. *Int. J. Hydrogen Energy*, 32, 3230–3237, 2007.
14. Dincer, I. and Acar, C., Smart energy solutions with hydrogen options. *Int. J. Hydrogen Energy*, 43, 18, 8579–8599, 2018.
15. Nastasi, B. and Lo Basso, G., Hydrogen to link heat and electricity in the transition towards future smart energy systems. *Energy*, 110, 5–22, 2016.
16. Europe Hydrogen Refuelling Infrastructure, Available online: https://h2me.eu/about/hydrogen-refuelling-infrastructure/ (accessed on 11 April 2019). (accessed on 11 April 2021).
17. USA Hydrogen Refuelling Infrastructure, Available online: https://afdc.energy.gov/fuels/hydrogen_locations.html#/find/nearest?fuel=HY (accessed on 11 April 2021).
18. Wei, Q., Zhang, X., Oh, B., The effect of driving cycles and H2 production pathways on the lifecycle analysis of hydrogen fuel cell vehicle: A case study in South Korea. *Int. J. Hydrogen Energy*, 46, 7622–7633, 2021.
19. Bethoux, O., Hydrogen fuel cell road vehicles: State of the art and perspectives. *Energies*, 13, 5843, 2020.
20. Bethoux, O., Hydrogen fuel cell road vehicles and their infrastructure: An option towards an environmentally friendly energy transition. *Energies*, 13, 6132, 2020.
21. Sapre, S., Pareek, K., Rohan, R., Singh, P., H2 refueling assessment of composite storage tank for fuel cell vehicle. *Int. J. Hydrogen Energy*, 44, 23699–23707, 2019.
22. Fonseca, J.D., Camargo, M., Commenge, J.-M., Falk, L., Gil, I.D., Trends in design of distributed energy systems using hydrogen as energy vector: A systematic literature review. *Int. J. Hydrogen Energy*, 44, 9486–9504, 2019.

23. Larminie, J. and Dicks, A., *Fuel cell systems explained*, 2nd ed, John Wiley & Sons, Ltd, West Sussex, UK, 2003.
24. Acar, C. and Bicer, Y., Review and evaluation of hydrogen production options for better environment. *J. Cleaner Prod.*, 218, 835–849, 2019.
25. Handwerker, M., Wellnitz, J., Marzbani, H., Comparison of hydrogen powertrains with the battery powered electric vehicle and investigation of small-scale local hydrogen production using renewable energy. *Hydrogen*, 2, 76–100, 2021.
26. Sunliang, C., Comparison of the energy and environmental impact by integrating a H_2 vehicle and an electric vehicle into a zero-energy building. *Energy Convers. Manage.*, 123, 153–173, 2016.
27. Koroma, M., Brown, N., Cardellini, G., Messagie, M., Prospective environmental impacts of passenger cars under different energy and steel production scenarios. *Energies*, 13, 6236, 2020.
28. Lys, A., Fadonougbo, J., Faisal, M., Suh, J., Lee, Y., Shim, J., Park, J., Cho, Y., Enhancing the hydrogen storage properties of AxBy intermetallic compounds by partial substitution: A short review. *Hydrogen*, 1, 38–63, 2020.
29. Tarkowski, R., Underground hydrogen storage: Characteristics and prospects. *Renewable Sustainable Energy Rev.*, 105, 86–94, 2019.
30. Aceves, S.M., Berry, G.D., Martinez-Frias, J., Espinosa-Loza, F., Vehicular storage of hydrogen in insulated pressure vessels. *Int. J. Hydrogen Energy*, 31, 2274–2283, 2006.
31. Stenmark, L., Uppsala University, Uppsala, Sweden, Personal communication, 2010.
32. European Union, Commission regulation (EU) No 406/2010 of 26 April 2010 implementing regulation (EC) No 79/2009 of the European Parliament and of the Council on type-approval of hydrogen-powered motor vehicle, Official Journal of the European Union.
33. Kim, Y.S., Kim, S.S., Choe, B.H., The role of hydrogen in hydrogen embrittlement of metals: The case of stainless steel. *Metals*, 9, 406, 2019.
34. Fu, L. and Fang, H., Formation criterion of hydrogen-induced cracking in steel based on fracture mechanics. *Metals*, 8, 940, 2018.
35. Adams, P., Bengaouer, A., Cariteau, B., Molkov, V., Venetsanos, A., Allowable hydrogen permeation rate from road vehicles. *Int. J. Hydrogen Energy*, 36, 2742–2749, 2011.
36. Stetson, N.T., Hydrogen storage overview, in: *Proceedings of the DoE Annual Merit Review and Peer Evaluation Meeting*, Washington, DC, USA, 16–20 June 2014.
37. Aigueperse, J., Mollar, P., Devilliers, D., Chemla, M., Faron, R., Romano, R., Cuer, J., Fluorine compounds, inorganic, in: *Ullmann's Encyclopedia of Industrial Chemistry*, vol. 15, Wiley, Hoboken, NJ, USA, 2012.
38. Hales, T. and Ferguson, S.A., Formulation of the kepler conjecture. *Discrete Comput. Geom.*, 36, 21–69, 2006.

39. Baptista, A., Pinho, C., Pinto, G., Ribeiro, L., Monteiro, J., Santos, T., Assessment of an innovative way to store hydrogen in vehicles. *Energies*, 12, 1762, 2019.
40. Hales, T., Adams, M., Bauer, G., Dang, T., Harrison, J., Hoang, L., Zumkeller, R.A., Formal proof of the kepler conjecture. *Forum Math.*, 5, 1–29, 2017.
41. Silbert, L., Jamming of frictional spheres and random loose packing. *Soft Matter*, 6, 2918–2924, 2010.
42. Dong, K., Yang, R., Zou, R., Yu, A., Role of interparticle forces in the formation of random loose packing. *Phys. Rev. Lett.*, 96, 145505, 2006.
43. Onoda, G. and Liniger, E., Random loose packings of uniform spheres and the dilatancy onset. *Phys. Rev. Lett.*, 64, 2727–2730, 1990.
44. Pouliquen, O., Nicolas, M., Weidman, P., Crystallization of non-brownian spheres under horizontal shaking. *Phys. Rev. Lett.*, 79, 3640–3643, 1997.
45. Donev, A., Improving the density of jammed disordered packings using ellipsoids. *Science*, 303, 990–993, 2004.
46. Man, W., Donev, A., Stillinger, F., Sullivan, M., Russel, W., Heeger, D., Chaikin, P., Experiments on random packings of ellipsoids. *Phys. Rev. Lett.*, 94, 198001-1–198001-4, 2005.
47. Scott, G. and Kilgour, D., The density of random close packing of spheres. *J. Phys. D Appl. Phys.*, 2, 863–869, 1969.
48. Scott, G., *Packing of spheres*, pp. 908–909, Nature Publishing Group, London, UK, 1960.
49. Barth, R.R., Simmons, K.L., San Marchi, C., *Polymers for hydrogen infrastructure and vehicle fuel systems: Applications, properties, and gap analysis*. Sandia National Laboratories Albuquerque, New Mexico and Livermore, California, 2013.
50. Perry, R.H., Green, D.W., Maloney, J.O., *Perry's chemical engineers' handbook*, 7th ed., McGraw-Hill, New York, USA, 1997.
51. Song, W., Du, J., Xu, Y., Long, B., A study of hydrogen permeation in aluminum alloy treated by various oxidation processes. *J. Nucl. Mater.*, 246, 139–143, 1997.
52. Van Deventer, E.H. and Maroni, V.A., Hydrogen permeation characteristics of some austenitic and nickel-base alloys. *J. Nucl. Mater.*, 92, 103–111, 1980.
53. Schefer, R.W., Hout, W.G., San Marchi, C., Chernicoff, W.P., Englom, L., Characterization of leaks from compressed hydrogen dispensing systems and related components. *Int. J. Hydrogen Energy*, 31, 1247–1260, 2006.
54. Paiva, L.B., Morales, A.R., Guimarães, T.R., Propriedades mecânicas de nanocompósitos de polipropileno e montmorilonita organofílica. *Polímeros*, 16, 2, 136–140, 2006.
55. Paul, D.R., Reformulation of the solution-diffusion theory of reverse osmosis. *J. Membr. Sci.*, 241, 371–386, 2004.
56. Humpenoder, J., Gas permeation of fibre reinforced plastics. *Cryogenics*, 38, 143–147, 1998.

57. Steward, S.A., *Review of hydrogen isotope permeability through materials*, Lawrence Livermore National Laboratory, Berkeley, California, USA, 1983.
58. Suda, H., Yamauchi, H., Uchimaru, Y., Fujiwara, I., Haraya, K., Preparation and gas permeation properties of silicon carbidebased inorganic membranes for hydrogen separation. *Desalination*, 193, 252–255, 2006.
59. Schultheiß, D., *Permeation barrier for lightweight liquid hydrogen tanks*, Phd Thesis, Universität Augsburg, Augsburg, Germany, April 16 2007.
60. Crowl, A. and Jo, Y., The hazards and risks of hydrogen. *J. Loss Prev. Process Ind.*, 20, 158–164, 2007.
61. Dwivedi, S.K. and Vishwakarma, M., Hydrogen embrittlement in different materials: A review. *Int. J. Hydrogen Energy*, 43, 21603–21616, 2018.
62. Echtermeyer, A.T., Lasn, K., Myers, B., *Safety factors and test methods for composite pressure vessels*, European Union, Brussels, Belgium, 2013.
63. Pinto, G., Monteiro, J., Baptista, A., Ribeiro, L., Leite, J., Study of the permeation flowrate of an innovative way to store hydrogen in vehicles. *Energies*, 14, 6299, 2021.
64. Waterland, L.R., Powars, C., Stickes, P., *Safety evaluation of the fuelmaker home refueling concept final report*, National Renewable Energy Laboratory, Golden, Colorado, USA, 2005.
65. Hyundai NEXO press kit, Available online: https://www.hyundai.news/eu/press-kits/0/NEXO/ (accessed on 19 February 2018).
66. Outline of the Mirai, Available online: https://www.toyota-europe.com/new-cars/mirai/index/specs (accessed on 19 February 2018).
67. Saffers, J.B., Makarov, D., Molkov, V.V., Modelling and numerical simulation of permeated hydrogen dispersion in a garage with adiabatic walls and still air. *Int. J. Hydrogen Energy*, 36, 2582–2588, 2011.
68. Venetsanos, A.G., Papanikolaou, E., Cariteau, B., Adams, P., Bengaouer, A., Hydrogen permeation from CGH2 vehicles in garages: CFD dispersion calculations and experimental validation. *Int. J. Hydrogen Energy*, 35, 8, 3848–3856, 2010.
69. Gentilhomme, O., Proust, C., Jamois, D., Tkatschenko, I., Cariteau, B., Studer, E., Masset, F., Joncquet, G., Amielh, M., Anselmet, F., Data for the evaluation of hydrogen risks onboard vehicles: Outcomes from the French project drive. *Int. J. Hydrogen Energy*, 37, 22, 17645–17654, 2012.

3

A Review on Hydrogen Compression Methods for Hydrogen Refuelling Stations

Nikolaos Chalkiadakis[1,2*], Athanasios Stubos[2], Emmanuel Stamatakis[3], Emmanuel Zoulias[4] and Theocharis Tsoutsos[1]

[1]*Renewable and Sustainable Energy Systems Lab, School of Chemical and Environmental Engineering, Technical University of Crete, Chania, Crete, Greece*
[2]*NCSR "DEMOKRITOS", Athens, Greece*
[3]*Institute of Geoenergy/Foundation for Research and Technology - Hellas (IG/FORTH), Chania, Greece*
[4]*New Energy & Environmental Solutions & Technologies - NEEST, Athens, Greece*

Abstract

Hydrogen is considered to be the fuel of the future, since it has the potential to store excess energy produced by renewable energy sources and then be used either as energy storage for power production by utilizing fuel cells or as a transportation fuel. For these reasons, it can be considered as the missing link between various industrial sectors, leading at the same time to increased renewable energy penetration and the decarbonisation of the sectors in which it is used. Since the transportation sector presents one of the most polluting sectors of human activity, it can be easily understood that utilizing hydrogen as a fuel for vehicles could lead to a significant reduction in carbon emissions. The main issues regarding the large-scale adoption of hydrogen as a fuel are its storage and its efficient compression. The solution of the latter issue, though, can lead to the solution of the former as well, since storing compressed hydrogen in high-pressure vessels presents one of the most commonly used methods of storage and the higher the pressure in which it is stored, the higher its energy content per unit volume. One could assume that by developing cost-efficient and environmentally friendly compression methods, the cost of compression, which currently amounts to almost half of the total cost of a hydrogen refuelling station, would drop substantially, leading to further adoption of hydrogen as a fuel for transportation. This work presents a review on state of the art for hydrogen compressors, be it either mechanical or non-mechanical,

**Corresponding author*: n.chalkiadakis@ipta.demokritos.gr

covering their operation principles, their applications as well as their benefits and drawbacks. More specifically, the contents of the current chapter area include: reciprocating, diaphragm, metal hydride and electrochemical compressors.

Keywords: Hydrogen, compressors, refuelling, energy storage

3.1 Introduction

Since the beginning of the 21st century the use of gaseous hydrogen as a fuel for various means of transportation has drawn the interest of scientists, governments, institutions and the automotive industry in general. One of the main arguments in favour of its adoption as the fuel of choice for the future, is the fact that it has the potential to connect various aspects of human activity such as power generation and storage, industry and transportation, leading to what has been called the "hydrogen economy". At the same time, it can provide an environmentally friendly solution for electric power-demanding uses, since it produces no greenhouse gas emissions when used.

Reading the above, one could wonder why hydrogen has not been widely accepted as a modern solution for problems which have troubled humanity for the past decades. While the public perception (having linked hydrogen with nuclear weapons and the Hindenburg disaster) regarding hydrogen focuses on safety issues, experts have identified the main barriers for its large-scale use, as the needs for its efficient storage and compression. Both issues have been examined extensively by scientists and there are already numerous methods to address them, at all levels of technical maturity.

Regarding its storage, hydrogen can be stored in 3 forms: gaseous, liquid and solid (metal hydrides). While in all three cases one can find benefits and drawbacks, the fact that gaseous hydrogen can be more flexible in terms of being integrated in a wider range of potential systems (such as the use as a fuel for automotive applications, injection into natural gas, etc.), makes it the most popular storage option. However, due to its low density, its storage only makes sense in high pressure, which achieves an increased energy density in terms of volume. This in turn leads to two main issues which have to be addressed: The identification of the best-suited materials which can tolerate high pressure hydrogen without being compromised, and the need for the development of energy efficient compressors, capable of reaching the desired pressures and volumetric flows.

In the following paragraphs, the reader can find a review of the most popular existing solutions regarding hydrogen compression, their pros and cons and their technical characteristics. The compressor types can be classified in two main categories: Mechanical and non-mechanical compressors. The former category consists of compressors which achieve compression by directly converting mechanical energy into compressed gas energy. They are currently the technology which can be found in the majority of hydrogen-related applications, since they can reach high pressures and achieve relatively high flow rates. The latter category, consists of compressors which make limited use of moving parts and instead achieve compression mainly by electro-chemical reactions. The main options regarding non-mechanical compression include metal hydride and electrochemical compressors, both of which are still in a pretty early stage of their development. However they show great potential, and their future adoption as the technologies of choice seems like a likely scenario.

3.2 Mechanical Compressors

Mechanical compressors are widely accepted as the technology of choice regarding gas compression for at least the last hundred years. This becomes evident by their current dominance in the gas compressor market. Mechanical compressors can be further divided in two main sub-categories, and more specifically to positive displacement and dynamic compressors [1]. For the purposes of this analysis, dynamic compressors will not be examined, since they present some significant challenges when it comes to compressing hydrogen. More specifically, centrifugal compressors in particular, due to hydrogen's low molecular weight, require high impeller tip speeds, which leads to a need for deployment of high-strength materials. This deems them techno-economically inferior, although currently there is extensive research conducted, in order to address this issue. The following paragraphs will be used for the analysis of positive displacement compressors, and in particular two of the most widely used types, which are piston and diaphragm reciprocating compressors.

3.2.1 Reciprocating Piston Compressors

Reciprocating piston compressors have been widely used for many different applications in which high pressure gases are required. They show great flexibility in terms of the discharge pressure which they can provide, as well as the respective flow rates. Also, they showcase relatively high efficiency, while they are capable of compressing gases of different densities and overall properties.

In order to have a complete understanding of how reciprocating piston compressors operate, one should delve deep into multiple scientific domains, including thermodynamic, mass and heat transfer, fluid dynamics, machine elements, etc. Since this is a task which escapes this chapter's purposes, the following paragraphs will be used for the analysis of the essential aspects of their operation, such as their basic components and their function as well as the most important equations which describe their operation and which can be used for calculating some of their key parameters.

3.2.1.1 *Basic Components and Operation of Reciprocating Piston Compressors*

Piston compressors can consist by either a single or multiple stages, depending on the specific needs of a given application. Regardless of their number, each stage includes the following components [2]: A piston, a connecting rod which connects the piston to a slider crank mechanism and is housed inside the cylinder, and two main valves facilitating hydrogen suction and discharge- a process which occurs automatically due to their design. Also used, are sealings, rings, liners, filters and additional auxiliary equipment. All the above can be seen in Figure 3.1 [3].

The operation of the compressor can be divided to three phases, which include suction, compression and discharge. Figure 3.2, shows in an easy to

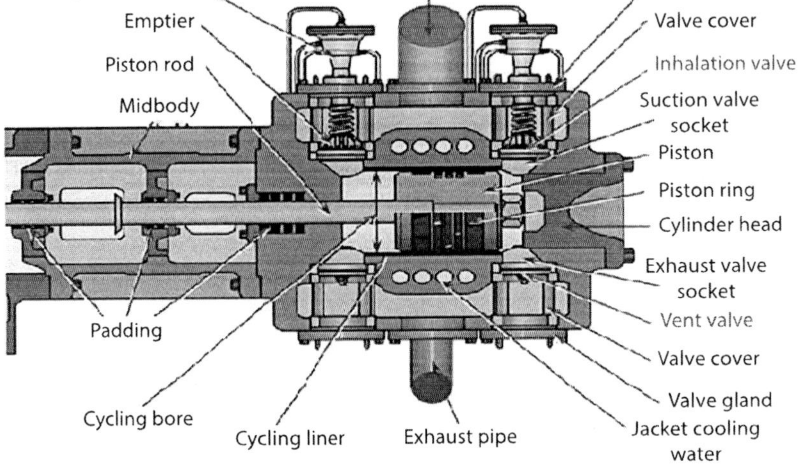

Figure 3.1 Components of a typical reciprocating piston compressor [3].

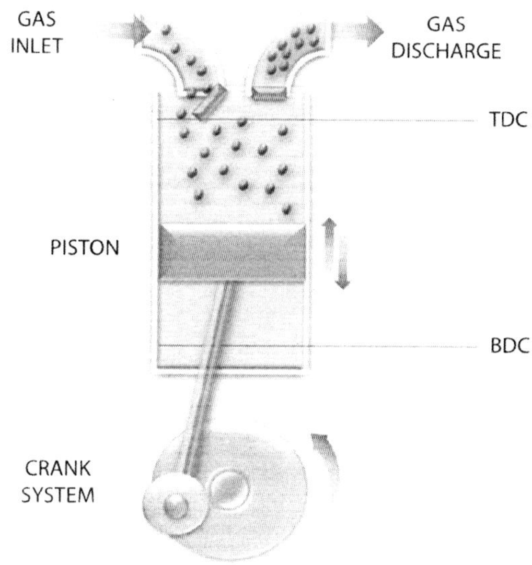

Figure 3.2 Visualization of the operation of a reciprocating compressor stage [2].

understand way, these three phases, in the context of the operation of a single stage, while introducing the concept of top dead center (TDC) and bottom dead center (BDC), which are the two extreme points between which the piston moves during its reciprocating motion. Basically, the way the above components work together to achieve compression is the following: Once the suction valve is open, hydrogen flows in the free volume between the piston and the cylinder head. Then the piston which is driven by the slider crank mechanism (which has the function of converting rotary motion to linear) moves towards the cylinder head, increasing hydrogen's pressure in the process. Afterwards, the discharge (or exhaust) valve opens and high pressure hydrogen leaves this compressor stage. What should also be noted, is the fact that the rotary motion of the slider crank, is usually achieved through the use of an electric motor which is connected to it. The process described will be analyzed in more detail in the following paragraphs.

3.2.1.2 Thermodynamic and Motion Dynamics Principles of Reciprocating Piston Compressors

As mentioned, the operation of reciprocating compressors is a multidisciplinary function, the most important aspects of which will be analyzed in the following paragraphs. To begin, the thermodynamic cycle of the reciprocating piston compressor will be presented.

3.2.1.2.1 Thermodynamic Cycle

The whole process of compression by reciprocating compressors, can be divided to four sub-processes which are determined in large by the relative position of the piston and the cylinder head, as well as the state of the suction and discharge valves. Diagram 3.1 [4] (known as Watt diagram) describes these four sub-processes, which in essence are the following:

- Process 1–2 describes the compression of hydrogen by the piston inside the cylinder, which is a result of the linear motion of the piston, made possible by the crankshaft's rotation.
- Process 2–3 describes the discharge of hydrogen to a discharge chamber, via the discharge valve. At point 2 the discharge valve opens, while the piston moves towards the TDC and at the same time, hydrogen which is at the design pressure passes through the discharge valve into the discharge chamber. Point 3 represents the TDC, where the discharge valve shuts and the part 2-2A-3 represents the discharging sequence.
- Process 3–4 describes the expansion of the gas volume which remained in the space between the piston and the cylinder head.
- Process 4–1 describes the suction of hydrogen from the stage's suction chamber into the cylinder. At point 4, the

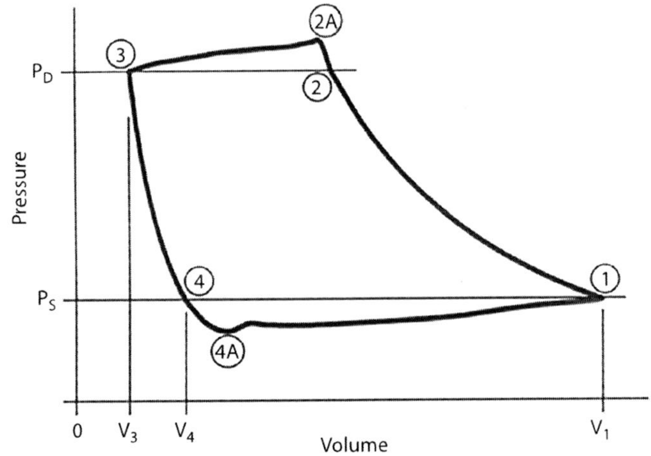

Diagram 3.1 Thermal cycle of a reciprocating piston compressor [4].

pressure is slightly lower than the suction pressure, resulting to the opening of the suction valve and the subsequent filling of the suction chamber. The part 4-4A-1 represents the suction sequence.

The work which is done in the above cycle, can be calculated by [5]:

$$W = \oint p\,dV \tag{3.1}$$

An important parameter for the assessment of any compressor stage, or any compressor in general is the compression ratio, which for a single stage, is defined as the discharge to suction pressure:

$$\Pi = \frac{p_{dis}}{p_{suc}} \tag{3.2}$$

For the case of multiple compression stages, the compression ratio is constant for all stages, and as a result the compression ratio for N number of stages, is:

$$\Pi_{Tot} = \Pi^{N} \tag{3.3}$$

Furthermore, the heat transfer rate from the environment to the fluid is given by:

$$\dot{Q} - W + \dot{m}_s h_s - \dot{m}_d h_d - \dot{m}_l h_l = E \tag{3.4}$$

where $m_s h_s$ and $m_d h_d$ are the rates of change of flow at the suction and discharge valves respectively and $m_l h_l$ is the rate of change of flow energy lost by leakage through the piston seals [4]. Additionally, in order to estimate a stage's efficiency, the isentropic efficiency factor (n_{is}) is used. The reasoning behind using this factor, stems from the need to take into consideration the fact that the processes which have been described are neither adiabatic nor reversible. Essentially, what this factor represents, is the degree of deviation of the compressor stage from the theoretically ideal value. Thus, the equation which is used is:

$$n_{is} = \frac{T_s}{T_d - T_s}\left[\left(\frac{P_d}{P_s}\right)^{\frac{k-1}{k}} - 1\right] \tag{3.5}$$

where T_s and T_d are the suction and discharge temperatures respectively and k can be calculated by:

$$k = \frac{C_p}{C_v} \tag{3.6}$$

which for the case of hydrogen has the value of 1.41.

3.2.1.2.2 Dynamics of Slider-Crank Mechanism

For the analysis of the dynamic phenomena that occur within a compressor stage, some key parameters which have to be calculated are the piston speed and the forces which result from its movement. Using Diagram 3.2, the following equations can be deducted [6].

The relative position as well as the speed of the piston is expressed as a function of the relative crank angle as well as of the dimensions of the crank and the connecting rod:

$$x(\theta) = -r(1 + cos\theta) + \sqrt{L^2 - r^2 sin\theta} \tag{3.7}$$

$$\frac{dx}{d\theta} = rsin\theta - r^2 sin\theta cos\theta / \sqrt{L^2 - r^2 sin\theta} \tag{3.8}$$

where r and θ are the crank radius and angle respectively and L is the connecting rod length.

The above equations finally lead to the deduction of the piston force F, which is a key parameter for the calculation of the structural integrity of the assembly

$$F_P = \frac{\tau\omega}{dx/d\theta} \tag{3.9}$$

where τ and ω are the shaft's torque and rotational speed respectively.

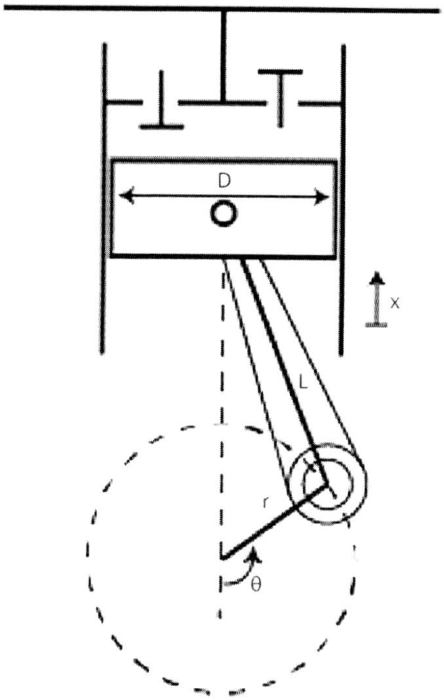

Diagram 3.2 Visualization of relative motions within a piston compressor stage [6].

3.2.2 Reciprocating Diaphragm Compressors

Due to the fact that diaphragm compressors achieve compression by a crank's reciprocating motion, they also fall into the category of reciprocating compressors. Their main differentiation from the reciprocating piston compressors, is the fact that the piston is not in direct contact with the compressed gas. Instead, the reciprocating motion of the piston is transferred through a hydraulic fluid to a diaphragm, which is also responsible for the isolation of the compressed gas. This separation between the two fluids, deems diaphragm compressors ideal for applications which require high gas purity, such as the use of hydrogen in fuel cells. Diaphragm compressors can have multiple stages, and are thus able to reach significantly high pressures, so they have been widely used for the purpose of hydrogen compression in hydrogen refueling stations. Some additional benefits of these compressors are the lack of need for cooling between stages, as well as their limited power requirements. On the other hand, the presence of the diaphragm assembly itself is a compromising factor, since it is the component which fails frequently, due to the high pressure gradients to which it is exposed. As a result there is the need for it frequent maintenance, which drives higher the total maintenance costs. Finally, another significant drawback which stems from the fact that the compressor is directly exposed to high pressure and high temperature hydrogen, is the issue of hydrogen embrittlement of its metallic surfaces.

3.2.2.1 *Reciprocating Diaphragm Compressor Components*

As mentioned, diaphragm compressors share the same components as piston compressors for the conversion of the rotary motion of the motor to the linear reciprocating motion of the piston. As a result components such as the crank mechanism, the piston, the cylinder, as well as the suction and discharge valves are common to both technologies. However, diaphragm compressors include some additional components which facilitate the use of the hydraulic fluid for the compression processes. As seen in Figure 3.3 [7], the uniformity of the pressure distribution in the diaphragm, is ensured by the use of a perforated plate which is in contact with the hydraulic fluid (oil). The hydraulic fluid is contained in an appropriate chamber which communicates with the cylinder, while the actual compression takes place in a chamber which is confined by the cavity surface and the diaphragm. A key design characteristic of the diaphragm assembly, is the geometry of this cavity surface, which as it will be explained, contacts the diaphragm when the piston is at its top dead center. The optimal design of this surface is of great importance, since it constitutes the primary cause of the stresses which are developed on the diaphragm, and as a result it can play a role in the integrity of the assembly.

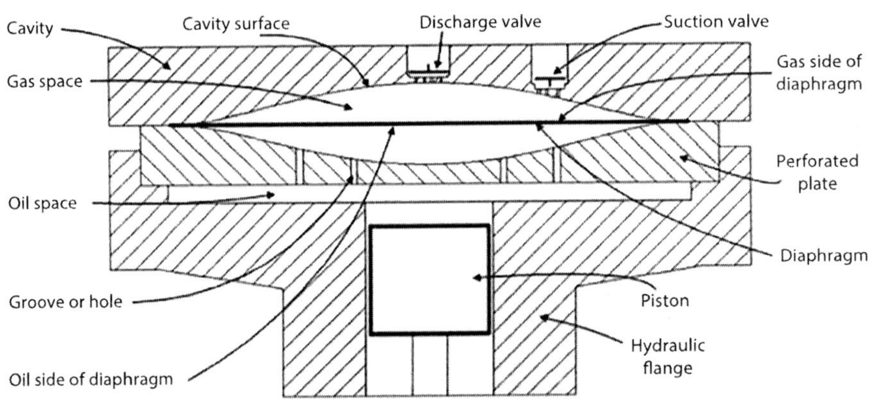

Figure 3.3 Components of a diaphragm compressor stage [7].

3.2.2.2 *Operating Principle of Diaphragm Compressor*

The operation of a diaphragm compressor stage is visualized in Figure 3.4. In detail, it can be described as follows. In the same manner as in piston compressors, the crank mechanism is driven by an outer source, which is usually an electric motor. This mechanism, as mentioned, has the function of converting the rotary motion of the motor, into the linear reciprocating motion of the piston. The piston, as a result, moves from the bottom dead center, towards the top dead center. When the position of the piston moves towards the bottom dead center, the diaphragm moves towards the perforated plate, and as a result the volume of the gas chamber increases. This leads to a slight drop in pressure, which is enough for the suction valve to open, and for the suction phase to begin. This phase lasts until the piston reaches the bottom dead center, where it stops momentarily, the suction valve closes, and the piston begins its motion towards the top dead center. While the piston moves towards the top dead center, the hydraulic fluid exerts a homogenous force on the diaphragm, which is pushed towards the cavity surface, compressing the gas which was in the gas chamber in the process. The discharge valve remains open until the desired pressure is reached and finally the diaphragm come in contact with the cavity surface, signaling the beginning of the next cycle.

While the thermodynamic equations that describe the function of the diaphragm compressor are practically the same as in the piston compressor, the equations for the structural and dynamic analysis are more complex and escape the nature of this chapter.

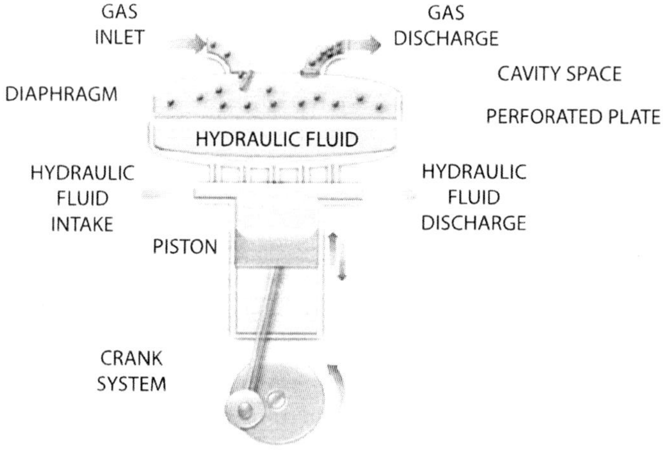

Figure 3.4 Visualization of the operation of a reciprocating compressor stage [2].

3.2.3 Integration of Reciprocating Piston Compressors in Hydrogen Refueling Stations

The main point which needs to be addressed for the integration of reciprocating compressors in hydrogen refueling stations, is the right selection in terms of operating pressure, flow rate and the electric motor's nominal power consumption. Obviously, the compressor should have the capability of achieving the desired hydrogen pressure, while having as input the low pressure hydrogen which will be supplied to their first stage. This input pressure, depends on the hydrogen production method, which for the case of electrolyzers (which are a key component in green hydrogen refueling stations) ranges from 8-30 bar. However, regardless of the hydrogen production method, the compressor should have a compression rate high enough to reach the pressure required from the vehicles which will be filled. At the same time, the compressor should have a flow rate that matches the respective flow rate of the hydrogen supply, in order to ensure: a) the optimal operation of the station (avoiding potential bottlenecks and mismatches) and b) the facilitation of a daily number of vehicles which the designer of the station has pre-specified as the station's capacity. Finally, what should be taken into account, is the station's capability of supplying the necessary power for the compressor's electric motor operation. Of course, this is not an issue for grid-connected installations, but could be a compromising factor for stand-alone stations, that operate on renewable energy sources. In the latter case, attention should be paid to the adequate sizing of the power source, and of any potential energy storage infrastructure, so the function of the compressor remains unhindered at all times.

3.3 Non-Mechanical Compressors

The compressors which are currently used at hydrogen fuelling stations are generally either reciprocating piston or diaphragm compressors, both of which are conventional mechanical compression systems. Although reciprocating compressors present a mature solution which can be used for the compression of the majority of gases, they comes with certain limitations. First of all, the extended use of moving parts not only leads to an increased manufacturing (due to the more complex design) and maintenance cost (due to the higher frequency of maintenance activities), but also to higher noise and vibration. Also, reciprocating compressors have the drawback of not being efficient at high flow rates, while the need for reducing mechanical stresses dictates the use of lower speeds, further restricting the allowable flow rates [8]. Most of the aforementioned flaws of reciprocating compressors are common to diaphragm compressors as well, with the additional issue of a potential diaphragm failure due to radial stress, related to diaphragm deflection [9].

Due to the above reasons, what becomes clear is the emerging need for identifying and developing different solutions for hydrogen compression, which will address the shortcomings of conventional mechanical compressors in an efficient way. Currently the most promising alternatives in hydrogen compression include metal hydride and electrochemical compressors, each of which will be analyzed in the following paragraphs.

3.3.1 Metal Hydride Compressors

Metal hydride compressors (MHCs) have received a great amount of attention the past years, since they have the potential of providing high pressure hydrogen, while firstly they consume significantly lower electrical power and secondly they present the capability of being integrated with either renewable energy systems or industrial processes which produce waste heat. At the same time, metal hydride compressors, having no need for moving parts, showcase a noiseless operation meaning they can be installed and operated in residential areas at all times. Of course the absence of moving parts results in a smaller chance of component failures and directly leads to a cutback in maintenance expenditures, as the required maintenance activities are not as frequent as in the case of their counterparts which utilize moving parts.

3.3.1.1 Principle of Operation

The operation of metal hydride compressors is based on the reversible hydrogenation and dehydrogenation cycles of metal hydrides, which are thermally driven processes. More specifically, it is based on the hydrogenation at relatively low temperature and pressure, and the subsequent dehydrogenation at a relatively higher temperature and pressure. As such, metal hydride compressors are classified as thermally driven compressors, achieving compression by heating and cooling, instead of the gas displacement which takes place in conventional mechanical compressors.

As mentioned, metal hydride compressors take advantage of the properties of certain hydride forming metals or alloys, or intermetallic compounds, and more specifically of the interaction between them and hydrogen, which results in the formation of metal hydrides. The most used materials are AB_5 (e.g. $LaNi_5$, $La_{0.5}Ce_{0.5}Ni_5$) and AB_2 (e.g. Zr-V-Mn-Nb) type alloys [10].

In order to understand how metal hydride compressors achieve compression, first the nature of the hydrogenation (absorption) and dehydrogenation (desorption) processes should be explained. Both reactions can happen efficiently only within a narrow range of temperatures and pressures, which are dictated by the metal hydride composition. As a result, the aim of a potential compressor designer, should be maintaining those temperatures, by providing the appropriate amount of heat to the vessel which contains the metal hydrides. An important element for the efficient design of a compressor stage, stems from the exothermic/endothermic nature of the absorption/desorption processes. For example, since absorption is an exothermic reaction, the designer should ensure that the produced heat should be removed from the metal hydride vessel in a continuous manner, in order to keep the temperature practically constant for as long as the hydrogenation cycle lasts. This is achieved by the external flow of a cooling medium, which removes the produced heat via the heat transfer which occurs in the vessel's wall. Of course the opposite is true for the case of desorption.

Basically, the reversible reaction which was mentioned above, can be described by:

$$M(s)+\frac{x}{2}H_2(g)\xleftrightarrow{abs/des}MH(s)+Q$$

where M is the metal (or alloy, or intermetallic compound), MH is the metal hydride and Q is the reaction heat, while (s) and (g) refer to the respective phase of the reactants. As discussed, the absorption reaction is exothermic, so for the forward reaction $Q<0$, and respectively for the backward $Q>0$. The equilibrium pressure for the above reaction is determined by Equation 3.10, which is the van't Hoff equation [11],

$$ln\frac{P}{P_o}=\frac{\Delta H}{RT}-\frac{\Delta S}{R} \tag{3.10}$$

where P is the hydrogen equilibrium pressure, P_0 is the standard pressure, ΔH and ΔS are the standard reaction enthalpy and entropy changes for hydride formation respectively.

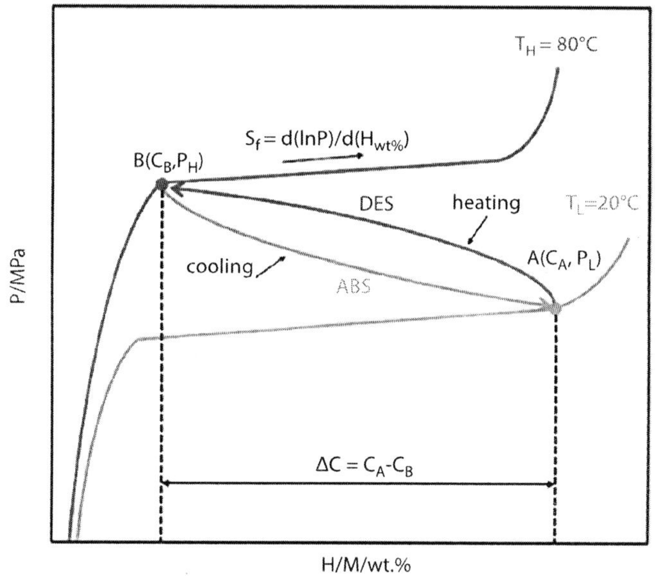

Diagram 3.3 Schematic representation of metal hydride absorption/desorption [12].

A schematic representation of the processes of absorption and desorption can be seen in Diagram 3.3 [12], which shows the P-C-T curves for hydrogen absorption and desorption.

In Diagram 3.3, one can identify the change in the metal hydride relative hydrogen concentration (C), in correlation to its pressure (P) and temperature (T). So, it can be understood that in this scenario, the metal hydride absorbs hydrogen at the equilibrium pressure (P_L) which corresponds to the temperature of 20 °C (T_L), until it becomes saturated (meaning it has reached it maximum hydrogen capacity (C_A)), and while being heated to 80 °C (T_H) it desorbs it, at the equilibrium pressure (P_H) until it reaches the respective concentration of C_B. Thus, we can conclude that the effective hydrogen compression capacity per cycle is:

$$\Delta C = C_A - C_B \tag{3.11}$$

3.3.2 Typical Metal Hydride Compressor Stage

Having discussed the basic aspects of the metal hydride absorption/desorption processes, it now makes sense to delve deeper into how a single stage of a metal hydride compressor operates. To do so, first some thermodynamic equations will be presented, in order to give the reader the tools for basic heat and mass transfer calculations. Then, some basic metal hydride compressor stage designs will be presented.

3.3.2.1 *Thermodynamic Analysis of Single Metal Hydride Compressor Stage*

Since metal hydride compressors are thermally driven machines, it makes sense to give an overview of the governing heat and mass transfer equations.

The following Equations 3.12 and 3.13 [13] describe the absorption kinetics.

$$m_g = -C_d \exp\left[-\frac{E_d}{RT}\right]\frac{P_g - P_{eq}}{P_{eq}}\rho_s \tag{3.12}$$

$$m_g = -C_a \exp\left[-\frac{E_a}{RT}\right] ln\left[\frac{P_g}{P_{eq}}\right](\rho_{ss} - \rho_s) \tag{3.13}$$

where m_g represents the hydrogen mass absorbed or desorbed (kg/m^{3}s), C_d and C_a (s^{-1}) represent the reaction rate constant for desorption and adsorption respectively, E_d and E_a (J/mol H_2) represent the activation energy for desorption and adsorption, while R, T, P and ρ represent the universal gas constant (J/molK), temperature (K), pressure (bar) and density (kg/m^3) respectively.

The energy balance equation for the case of a cylindrical metal hydride vessel can be estimated by Equation 3.14 [14].

$$(\rho C_p)_e \frac{\partial T}{\partial t} = \frac{1}{r}\frac{\partial}{\partial r}\left(\frac{rk_e \partial T_g}{\partial r}\right) + \frac{\partial}{\partial z}\left(\frac{k_e \partial T_g}{\partial z}\right) - (\rho C_{pg} V_{gr})\frac{\partial T}{\partial r}$$
$$-(\rho C_{pg} V_{gz})\frac{\partial T}{\partial z} - m\left(\frac{\Delta H}{M_g} - T(C_{pg} - C_{ps})\right) \tag{3.14}$$

where C_p is the specific heat (J/kgK), k is the thermal conductivity (W/mK), t is time (s) and M molecular weight (kg/mol), while r and z refer to the radial and axial space coordinates (m). The subscripts g and s refer to the gaseous and solid phase, while e hints to the use of the effective heat capacity and effective thermal conductivity which are given by Equations 3.15 and 3.16.

$$(\rho C_p)_e = \varepsilon \rho_g C_{pg} + (1 - \varepsilon)\rho_s C_{ps} \tag{3.15}$$

$$k_e = \varepsilon k_g + (1 - e)k_s \tag{3.16}$$

Hydrogen Mass Balance
Regarding the hydrogen mass balance, the continuity equation is used, again taking into consideration a cylindrical hydrogen vessel. Thus the mass balance of hydrogen is given by Equations 3.17 and 3.18.

$$\varepsilon \frac{\partial \rho_g}{\partial t} + \vec{\nabla} \cdot \left(\rho_g \vec{V_g} \right) = -m \tag{3.17}$$

$$\rho_g = \frac{P_g M_g}{ZRT_g} \tag{3.18}$$

where Z is the compressibility factor and V_g is the hydrogen gas velocity, which in the case of porous media such as metal hydrides, is given by Equation 3.19, which is Darcy's equation.

$$\vec{V_g} = -\frac{K}{\mu_g} \vec{\nabla} P_g \tag{3.19}$$

where μ is the dynamic viscosity (kg/ms) and K is the porous media permeability (m^2).

For the case of the mass balance in the solid phase, Equation 3.20 is used.

$$(1-\varepsilon)\frac{\partial \rho}{\partial t} = m_s \tag{3.20}$$

The equations which were presented above, constitute the basis for the numerical analysis and the design of a single metal hydride compressor's stage.

3.3.2.2 *Metal Hydride Compressor Stage Design*

The design of a metal hydride stage should on the one hand take into consideration the thermodynamic properties of the system which were described in the previous paragraph, and on the other solve a number of practical problems such as its optimal heating and cooling and its correct sizing.

By having knowledge of the exact metal hydride composition and its characteristics, one has the ability to estimate the temperature and pressure range within which it will operate in the context of a compressor stage. As a result, the amount of heat which will be needed for reaching the desired temperatures can be estimated. Attention should be paid to the exothermic/endothermic nature of the reactions, which present an additional thermal load which needs to be considered and is directly related to the mass of the metal hydride which is present in the stage. Also important, is knowledge of the specific heat capacity of the vessel's wall material, as well as its thickness, since it is in direct contact with the metal hydrides, and as such it should also be heated/cooled [15]. Of utmost importance, of course, is ensuring that the hydride vessel can tolerate the pressure to which hydrogen is desorbed at, without suffering structural damage. Finally, the fact that during absorption, the volume of the metal hydride increases, leads to the conclusion that the hydride containing vessel should have an adequately large empty volume, in order for this expansion to be tolerated.

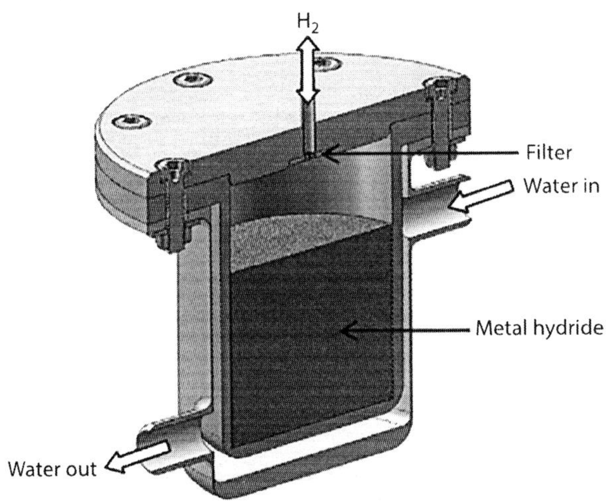

Figure 3.5 A metal hydride compressor stage utilizing an outer shell heating/cooling system [16].

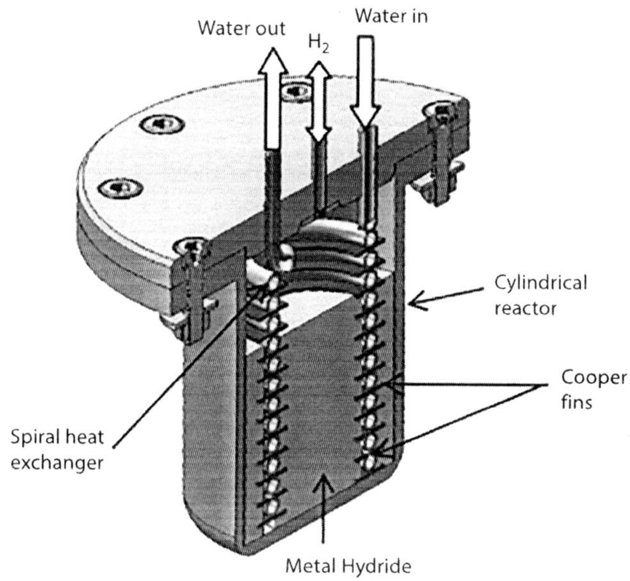

Figure 3.6 A metal hydride compressor stage utilizing a spiral finned heat exchanger for heating/cooling [16].

In order to provide the necessary heating and cooling loads to the stage, it is obvious that an external medium should be used. The selection of the medium should take in account the temperature which needs to be reached, so, for example, if the desired desorption temperature of a given metal hydride is above 100°C, it wouldn't make sense to use water as the heating medium. The flow of this external medium should be such, that it can achieve a uniformity in temperature in the metal hydride vessel. Attention should be paid not only to the flow rate, but to the method which is used for the heat exchange as well. A number of different ways have been applied, utilizing setups such as outer jackets (Figure 3.5) [16], tubes, spiral and finned heat exchangers, or their combinations (Figure 3.6) [16].

3.3.3 Metal Hydride Compressors Stages Integration

Achieving high compression rates is possible for metal hydride compressors only when numerous stages are employed. It is clear, that the only way this can be possible is through the selection of the right metal hydrides. To be specific, the design should ensure that the output pressure of each stage, matches the equilibrium pressure for the absorption of the next. Diagram 3.4 [17] is used to visualize the interconnection between two

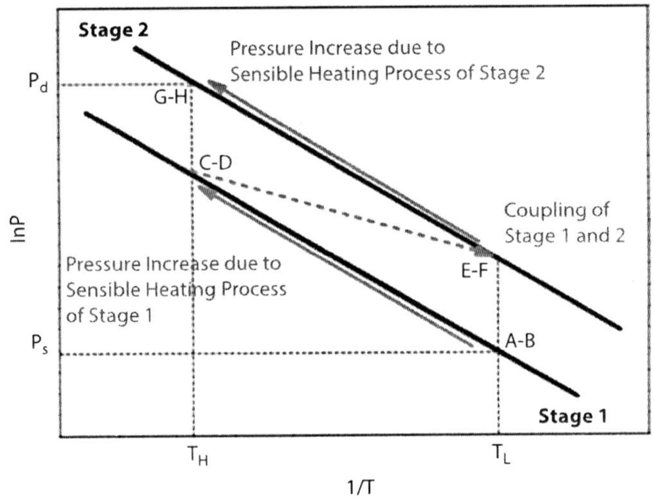

Diagram 3.4 Illustration of the operation of a two staged metal hydride compressor [17].

stages. The black lines represent the Van't Hoff plot for the hydrogenation process for stages 1 (lower black line) and 2 (upper black line). The coupling of the stages is represented by the dashed line.

3.3.4 Metal Hydride Compressor Integration in Hydrogen Refuelling Stations

The integration of a metal hydride compressor, in a complete system such as a hydrogen refuelling station, is a task which requires attention in a few key points, the most obvious of which is the compressor's capability of reaching the desired pressure (dictated by the installation needs) and flow rate. Currently, for stations which facilitate heavy duty vehicles, the desired pressure is 350 bar, while for ordinary passenger vehicles the respective pressure is 700 bar. Meanwhile, the flow rate of the compressor is dictated by the absolute number of vehicles that the station aims to facilitate on a daily/weekly basis.

Another aspect if the system which needs to be addressed, which is also relevant to the flow rate which is achieved by the compressor, is the optimal matching between system components such as the hydrogen production unit and the compressor. It is clear, that especially for green hydrogen refueling stations, which tend to deploy electrolyzers for the production of hydrogen, there is an upper limit to the hydrogen quantity that is produced per unit of time. As a result, the compression flow rate should match the electrolyzer output rate, in order to avoid either potential bottlenecks (meaning that the electrolyzer produces more hydrogen than the capacity of the compressor, thus leading to the decrease in electrolyzer efficiency), or the constant operation of the compressor at a capacity lower than its nominal one. For the case of the potential bottleneck, a simple solution would be utilizing a buffer storage space, which would receive the excess hydrogen, even though depending on the extent of the mismatch, it might just postpone the creation of the bottleneck.

What also need to be considered during the selection of the hydrogen producing unit (regardless of the compression type) is matching the pressure at which hydrogen is produced, with the suction pressure of the first stage of the compressor.

Finally, the most important aspect for the integration of the metal hydride compressor, is the availability of an adequate amount of the heating/cooling medium, ensuring its uninterrupted operation at all times [18]. This is accomplished by calculating the medium flow rate and the temperature ranges that are required for the metal hydride compressor operation. By estimating these values, one can then estimate the heat flow which is required at all times and then select the method with which this heat flow will be made available to the system. Some choices include using solar heaters/chillers, using resistances and heat pumps, or utilizing excess waste heat from industrial processes. The latter method makes the most sense in terms of economics and sector coupling, as it would provide benefits to both the industry which rejects heat and the metal hydride compressor operator, since there will be an abundance of low grade thermal energy available. Their capability of working with renewable energy sources and excess heat, actually proves to be one of their greatest advantages, while additional benefits can be considered their noiseless operation, the lack of CRM use and the minimal operation and maintenance costs. Their main drawbacks are their relatively immature technology, in terms of market readiness as well as their relatively low efficiency (although this can be of no impact in case an abundant waste heat source is available).

3.4 Electrochemical Compressors

3.4.1 Components and Operation of Electrochemical Compressors

Electrochemical hydrogen compressors (ECC) have become a viable solution when it comes to high efficiency purification and compression of hydrogen, especially for low pressure and low power-demanding applications. The reason behind their high efficiency at low power, stems from their operating principle (which will be analysed shortly), which leads to minimal irreversible energy losses at low current densities. It is for these applications, that they also showcase rather low capital and operating costs [2].

Electrochemical compressors have many similarities in their design with proton exchange membrane (PEM) fuel cells. The core of any ECC is an electrochemical cell, which is a membrane that acts as an electrolyte for protons exchange, positioned between anode and cathode electrodes. This setup basically forms the membrane electrode assembly, which is shown in Figure 3.7. In the same figure, one can identify the gaskets, gas diffusion layers, bipolar plates with gas flow channels, current collectors and end plates, as well as additional components of the ECC. The bipolar plates' gas flow channels ensure the homogenous transmission of the reactants to the gas diffusion layer. The porous gas diffusion layer is in contact with the anode and cathode catalyst layers, while at the same time they play a critical role in water and heat management [19]. The role of gaskets is purely of mechanical nature, preventing any potential leakage.

The basic function of the electrochemical cell (which is visualized in Figure 3.8) begins when an external voltage is applied across its electrodes. Once hydrogen enters the cell at the anode, it is oxidized, generating

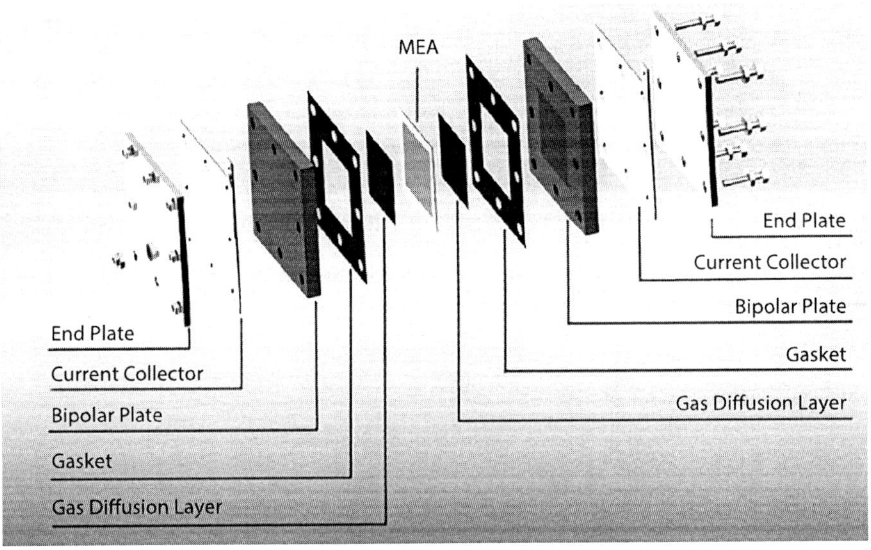

Figure 3.7 Components of electrochemical compressors [19].

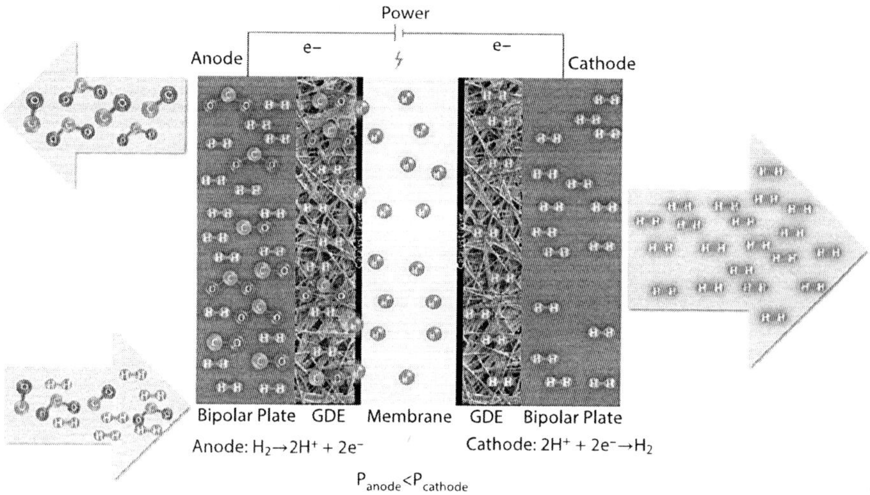

Figure 3.8 Visualization of the operating principle of electrochemical compressors [19].

protons and electrons in the process. Due to the membrane's properties, the protons can pass through it, in order to reach the cathode. The produced electrons also are directed to the cathode, via an external circuit, where they recombine with the protons, once again forming hydrogen at a higher pressure, due to the local accumulation of hydrogen molecules, which is a direct result of the existence of a backpressure regulator.

The reactions that take place at the anode and the cathode are:

$$H_2 \rightarrow 2H^+ + 2e^- \ (anode)$$

$$2H^+ + 2e^- \rightarrow H_2 \ (cathode)$$

In the first of the two equations, hydrogen is at relatively low pressure, while the amount of hydrogen which is produced in the second equation is at relatively high pressure.

The relationship between the applied current and the theoretical number of compressed hydrogen moles is [20]:

$$n_{theoretical} = \frac{I}{2F} \tag{3.21}$$

where I is the applied current and F the Faraday constant. The theoretical cell voltage is given by Nernst's equation as follows:

$$E_{Nernst} = \frac{RT}{2F} \ln \frac{P_C}{P_a} \tag{3.22}$$

where P_c and P_a refer to the cathode and anode pressure respectively. Taking into consideration additional factors such irreversibilities due to the activation overpotential and ohmic overpotential, the above equation is updated by:

$$E_{cell} = E_{Nernst} + \Delta E_{Activation} + \Delta E_{Ohmic} \tag{3.23}$$

where the overpotentials at the right part of the equation can be calculated by Equations 3.24–3.27.

$$\Delta E_{Activation} = E_{Activation\ anode} + E_{Activation\ Cathode} \tag{3.24}$$

$$E_{Activation\ anode} = \frac{RT}{2a_{anode}F} \ln\left(\frac{i}{i_{o,anode}}\right) \tag{3.25}$$

$$E_{Activation\ cathode} = \frac{RT}{2a_{cathode}F} \ln\left(\frac{i}{i_{o,cathode}}\right) \tag{3.26}$$

$$\Delta E_{Ohmic} = iR_{Ohmic} = i\frac{t_m}{\sigma} \tag{3.27}$$

where i is the current density (A/cm^2), i_0 is the exchange current density (A/cm^2), α is the charge transfer coefficient, R_{Ohmic} is the cell ohmic resistance, t_m is the membrane thickness (cm) and σ is the ionic conductivity of the cell (S/cm) [19].

What should also be considered for calculations, are mass transfer limitations, which prevail primarily at high current densities. These limitations result from the inability of the reacting gas to be transported to the anode quickly enough. Additionally, a key design point for ECC is the optimization of the membrane thickness. Reducing the membrane's thickness leads to a decrease in the ohmic overpotential, but attention should be paid to the minimum acceptable membrane thickness which can tolerate the pressure which is reached in the electrochemical cell. Another reason for the need for optimizing the membrane's thickness is the back diffusion which can occur from the cathode to the anode side of the membrane. Since pressure at the cathode is higher than in the anode, hydrogen is forced to diffuse towards the anode, preventing the system from reaching higher pressure. Back diffusion can be quantified via Fick's law as shown in Equation 4.8.

$$\dot{n}_{back\ diffusion} = DA\frac{\Delta C}{t_m} = DAH\frac{P_c - P_a}{t_m} \tag{3.28}$$

where D is the diffusivity of hydrogen through the membrane, A is the active area of the membrane, ΔC is the difference in the concentration of hydrogen between the two sides of the membrane, and H is the Henry constant [19]. The total power required for the compression of hydrogen by an ECC is given by Equation 4.9.

$$P = E_{Cell}I \tag{3.29}$$

A significant advantage of ECC stems from the absence of moving parts, which reduces their noise levels, as well as their operation and maintenance costs to significantly lower levels than the respective values of mechanical compressors. Additionally, the pressure increase takes place at a constant temperature, which removes the need for any intermediate cooling stages, leading to a lower capital cost [21]. Finally, ECCs are an energy efficient solution, especially for low power applications, while an additional benefit they showcase is the fact that they produce high purity hydrogen, due to their working principle. However, the challenge of developing a reliable membrane which not only has the necessary durability, but also solves the issue of back diffusion effectively is what keeps ECCs from being widely adopted, even though they present an interesting technology with great potential.

3.4.2 Integration of Electrochemical Compression in a Hydrogen Refuelling Station

As discussed in all previous segments regarding the integration of the compressor with the rest of the components of the hydrogen refuelling station, attention should be paid to matching the flow rate of the hydrogen production method with the compressor's capacity, in order to ensure the optimal operation of the installation. Also for the case of electrochemical compressors, there exists the need of having the capability to provide the necessary electric power for its operation. While this is not an issue for grid-connected installation, it is an important factor for stand-alone stations. It is of course evident, that the compressor should at all times be able to provide high enough pressure, for the operation of the vehicles which will be refilled in the station in question.

References

1. Don, W.G. and Robert, H.P., *Perry's chemical engineers' handbook*, Eighth Edition, 8th ed., McGraw-Hill Education, New York, 2008.
2. Sdanghi, G., Maranzana, G., Celzard, A., Fierro, V., Review of the current technologies and performances of hydrogen compression for stationary and automotive applications. *Renew. Sustain. Energy Rev.*, 102, 150–170, 2019.
3. Han, Y., Wang, J., Liu, H., Zhang, Z., The common faults and analysis of hydrogen compressor. *IOP Conf. Series: Earth Environ. Sci.*, 440, 2, 2020.
4. Hollingsworth, J., Phillippi, G., Hinchliff, M., Kulhanek, C., Rimpel, A., Maywald, F., Reciprocating compressors, in: *Compression Machinery for Oil and Gas*, pp. 167–252, 2019.
5. Stouffs, P., Tazerout, M., Wauters, P., Thermodynamic analysis of reciprocating compressors. *Int. J. Therm. Sci.*, 40, 1, 52–66, 2001.
6. Mohammadi-Amin, M., Jahangiri, A.R., Bustanchy, M., Thermodynamic modeling CFD analysis and parametric study of a near-isothermal reciprocating compressor. *Therm. Sci. Eng. Prog.*, 19, 3, 2020, Article 100624.
7. Jia, X., Zhao, Y., Chen, J., Peng, X., Research on the flow rate and diaphragm movement in a diaphragm compressor for a hydrogen refueling station. *Int. J. Hydrogen Energy*, 90, 41, 33, 14842–14851, 2016.
8. Almasi, A., Reciprocating compressor optimum design and manufacturing with respect to performance, reliability and cost. *World Acad. Sci. Eng. Technol.*, 52, 48–53, 2009.
9. Jiahao, C., Xiaohan, J., Chuang, X. *et al.*, Design and validation of new cavity profiles for diaphragm stress reduction in a diaphragm compressor. *IOP Conf. Ser. Mater. Sci. Eng.*, 2015.
10. Stamatakis, E., Zoulias, E., Tzamalis, G., Massina, Z., Analytis, V., Christodoulou, C., Stubos, A., Metal hydride hydrogen compressors: Current developments & early markets. *Renewable Energy*, 127, 850–862, 2018.
11. Goshome, K., Endo, N., Maeda, T., Demonstration of a single-stage metal hydride hydrogen compressor composed of BCC $V_{40}TiCr$ alloy. *Int. J. Hydrogen Energy*, 46, 55, 28180–28190, 2021.
12. Peng, Z., Li, Q., Ouyang, L., Jiang, W., Chen, K., Wang, H., Liu, J., Zhu, M., Overview of hydrogen compression materials based on a three-stage metal hydride hydrogen compressor. *J. Alloys Compd.*, 895, Part 1, art. no. 162465, 2022.

13. Muthukumar, P., Patel, K.S., Sachan, P., Singhal, N., Computational study on metal hydride based three-stage. *Int. J. Hydrogen Energy*, 37, 4, 3797–3806, 2012.
14. Gkanas, E., Christodoulou, C., Tzamalis, G., Stamatakis, E., Deligiannis, K., Karagiorgis, G., Stubos, A., Numerical investigation of a seven-stage metal hydride hydrogen compression system. *Renewable Energy*, 147-1, 164–178, 2020.
15. Gkanas, E., Stamatakis, E., Christodoulou, C., Tzamalis, G., Karagiorgis, G., Chroneos, A., Kuganathan, A., Khzouz, M., Stubos, A., Study on the operation and energy demand of dual-stage metal hydride hydrogen compressors under effective thermal management. *Int. J. Hydrogen Energy*, 46, 29272–29287, 2021.
16. Dhaou, H., Souahlia, A., Mellouli, S., Askri, F., Jemni, A., Ben Nasrallah, S., Experimental study of a metal hydride vessel based on a finned spiral heat exchanger. *Int. J. Hydrogen Energy*, 35, 4, 1674–1680, 2010.
17. Gkanas, E.I., Grant, D.M., Stuart, A.D., Eastwick, C.N., Book, D., Nayebossadri, S., Pickering, L., Walker, G.S., Numerical study on a two-stage metal hydride hydrogen compression system. *J. Alloys Compd.*, 645, S1, art. no. 33734, S18–S22, 2015.
18. Chalkiadakis, N., Stubos, A., Zoulias, E., Stamatakis, E., Pilot autonomous hybrid hydrogen refueling station utilizing a metal hydride compressor covering local transportation needs. *E3S Web of Conferences*, vol. 334, p. 06002, 2022.
19. Durmus, G.N.B., Colpan, C.O., Devrim, Y., A review on the development of the electrochemical hydrogen compressors. *J. Power Sources*, 494, art. no. 229743, 6, 2021.
20. Bampaou, M., Panopoulos, K.D., Papadopoulos, A.I., Seferlis, P., Voutetakis, S., An electrochemical hydrogen compression model. *Chem. Eng. Trans.*, 70, 1213–1218, 2018.
21. Toghyani, S., Baniasadi, E., Afshari, E., Javani, N., Performance analysis and exergoeconomic assessment of a proton exchange membrane compressor for electrochemical hydrogen storage. *Int. J. Hydrogen Energy*, 45, 60, 34993–35005, 2020.

4

Current Technologies and Future Trends of Hydrogen Propulsion Systems in Hybrid Small Unmanned Aerial Vehicles

Hasan Çınar[1], Ilyas Kandemir[2] and Teresa Donateo[3]*

[1]*Faculty of Aeronautics and Astronautics, Iskenderun Technical University, Iskenderun, Turkey*
[2]*Mechanical Engineering, Faculty of Engineering, Gebze Technical University, Gebze, Turkey*
[3]*Department of Engineering for Innovation, University of Salento, Lecce, Italy*

Abstract
The use of small unmanned aerial vehicles (UAVs) with electrical propulsion systems in both military and civilian areas has attracted considerable attention in recent years. In such UAVs, batteries, supercapacitors, hydrogen-fuel cell systems, piezo materials, and solar cells can be used to power the electric motors during the flight. These systems have different specifications in terms of cost, energy and power density, dynamic response, efficiency, lifetime, etc. and none of these energy sources alone can hardly meet the power demand of a UAV. For this reason, hybrid systems, which are a combination of these energy components, have been proposed in the literature. Such hybrid systems require appropriate energy management to increase the total efficiency of the system, lifetime of each energy component, and flight time/range of UAV. Energy management techniques applied to hybrid systems are divided into two classes, active or passive. Active energy management strategies are more efficient but require more complex system topologies than passive methods. This study presents different hydrogen-based systems that can be used to feed the electric motors in UAVs and compare them with other technologies in terms of several Key Performance Indexes. Moreover, different hybrid system topologies proposed for fuel cell-based small UAVs and their passive/active energy management techniques such as rule-based, optimization-based, and learning-based strategies are examined and discussed. Finally, future work

**Corresponding author*: teresa.donateo@unisalento.it

topics on the use of fuel cells in UAVs are proposed, which could be important to optimize the flight time/range and the energy consumption of UAVs.

Keywords: Fuel cell-based hybrid propulsion, energy/power management, unmanned aerial vehicles

4.1 Introduction of Fuel Cell-Based Propulsion for UAVs

In the field of transportation, there is a great interest to fully electric propulsion systems in aircraft and unmanned aerial vehicles (UAVs), as well as in ground vehicles, trains, ferries, etc. The reason is that engines dependent on fossil fuels are not environmentally friendly, vibrate, have high thermal traces, harmful emissions, and low efficiency.

Unmanned aerial vehicles (UAVs) are classified according to many factors including size, endurance, and flight altitude [1]. With respect to weight, those with a total mass of less than 20 kg are classified as small UAVs in [2] even when there is no generality in the usage of nano, micro, mini, light, small and tactical. For endurance and altitude, the acronyms MALE and HALE are used to denote "Medium Altitude Long Endurance" and "High Altitude Long Endurance", respectively. The use of small UAVs has increased in both the military and civilian areas in recent years. Today, they are used in many areas such as crop control and spraying in agriculture, border security, data collection in disaster situations, rescue missions, surveillance, taking pictures and videos, and various security applications [3–7]. In addition, small UAVs have been used recently in the early diagnosis and control of COVID 19 outbreak [8–10]. In this context, it is anticipated that the delivery of substances such as medicine and food by UAVs will increase in the near future [11, 12]. Other usage areas and mission purposes of UAVs are only limited by imagination.

Except for very few applications, the main goal in the design of the propulsion system of a UAV is to ensure the highest endurance (or range). In this content, the synergic optimization of the power system, vehicle design and mission specification can play an important role in increasing the flight time in small UAVs [13].

This chapter concentrates on the power system of UAVs which require less than 20 kW. For this application, an engine-based power system is not suitable [14]. In fact, small UAVs are mostly equipped with propellers connected to electric motors (that play the role of internal combustion engines used in larger UAVs). What differentiates the architectures for small UAVs proposed in literature, is how the electric motors are fed, i.e., the source of electricity used on board. This electricity may come directly from energy carriers like batteries or supercapacitors or converted from another form of energy, like in the case of hydrogen-powered UAVs where a fuel cell is used to convert the chemical energy of hydrogen into electricity. Similarly, a photo-voltaic panel can be used to convert solar energy into electric energy. Actually, many authors in the literature make confusion between energy sources (battery, supercapacitor, hydrogen storage, etc.) and converters (fuel cell, photo-voltaic panels, and so on) [15]. So, one of the goals of this chapter is to classify the components of electric UAVs in terms of energy sources, energy converters, and energy transformers.

These sources have different characteristics from each other in terms of energy/power density, dynamic response speed, service life, and energy weight ratio. The energy density refers to the energy stored per unit mass or volume and is directly related to the maximum endurance that can be obtained. The power density refers to the power that can be drawn from the energy source per unit mass and it is an important parameter because the UAVs may need high power during take-off, climb, and maneuvering. Energy and power density are particularly important to determine the applicability of an energy source to a UAV (see [16, 17]).

Supercapacitors have a very low energy density, so they cannot be used alone to feed the electric motors but are suitable to meet the peak loads [18] in the power demand. Solar cells, on the other hand, are not suitable for any platform because they require a large surface area. In addition, their ability to produce energy is limited by the presence of sunlight. Batteries can be used alone to meet the energy needs of UAVs, but since they have limited energy densities, they limit the flight time of the UAV. Lithium-polymer batteries are preferred in UAVs because their power density is higher than other types of batteries [19]. The maximum flight time for a lithium battery-driven UAV was found to be 90 minutes in the literature [20].

In UAVs, fuel cell systems are a good option because of the very high energy density of hydrogen [19]. In addition, their emissions are almost harmless to the environment [21]. The biggest obstacles for fuel cells to provide energy alone in UAVs is their low dynamic response rates that make them unsuitable to handle highly dynamic loads [22, 23]. Also, fuel cells have a complex and expensive and require accurate control strategies compared to other energy sources [24]. However, technological developments are expected to solve these drawbacks so that they will be used alone for the propulsion of UAVs in the future. Proton exchange membrane fuel cell (PEMFC), direct methanol fuel cell (DMFC), and solid oxide fuel cell

(SOFC) are commonly used types in UAVs, regardless of the class of the UAV [25]. In mini and micro class UAVs, mostly PEMFC and DMFC type fuel cells are used rather than SOFCs due to high operating temperature and cooling needs. DMFC fuel cells have low energy density and system efficiency compared to PEMFCs. For these reasons, PEMFC is the most suitable system for mini and micro class UAVs in terms of system efficiency, operating temperature, and energy density. To complete the analysis of the possible energy sources, one could cite the piezo materials. However, they produce very low amounts of power and energy [26] so they can be used to provide energy for electronic components rather than supplying energy to the drive system.

Since none of the energy sources described above can meet alone the power demand of a UAV, except for short flight times, hybrid solutions have been proposed in the literature. There are several studies in the literature on hybrid systems consisting of battery, fuel cell, and supercapacitor (e.g., [27–29]), a fuel cell and a battery (e.g., [30–35]), solar cell, battery, and fuel cell (e.g., [33–40]), solar cell and battery combinations (e.g., [41–43]). For some of these hybrid systems, successful flights have been carried out for more than 24 hours ([35, 38]). Such hybrid systems require suitable energy management strategies to define the usage of each energy source during the flight and improve the performance of the UAV. For example, in PEMFC hybrid systems with batteries and/or supercapacitors [43], suitable energy management techniques are needed to maintain total system efficiency, use energy sources in optimum operation, and protect the life of each source [44]. Moreover, real-time applicability, environmental issues, and other aspects like water/heat/gas management [45] need to be addressed.

Energy management is an important research topic in fuel cell-based hybrid propulsion systems of UAVs and intensive studies have been carried out in recent years. In addition, the conceptual design and size optimization of fuel cell propulsion systems including hydrogen storage are topics that need further research. In the first part of this chapter, the basic components of an electric UAV are classified and compared with appropriate key performance indexes. In the second part, the different technologies available for fuel cells are mentioned and PEMFCs are analyzed in detail. Then, the possible hybrid architectures with fuel cells are described. This is followed by a detailed analysis of the experimental and simulation studies of the fuel cell in the propulsion system of UAVs. Then, the specific topic of energy management in hybrid UAVs with fuel cells is addressed. Finally, possible future research topics of UAVs with fuel cells are proposed.

4.2 Unified Classification of the Components | of a Hybrid Electric Power System in UAVs

In aviation as in other transportation areas, the use of electric motors instead of internal combustion engines for propulsion provides many advantages such as silent and vibration-free operation, low thermal trace, high conversion efficiency, and absence of harmful gas emissions [46]. Like in any power system, we can identify a series of possible energy domains and classify each component of an electric-powered UAV in converters, transformers, and storage systems or energy sources [47]. In converting energy, two types of variables are important: flow variables (e.g., current, translational/angular velocity, and fuel flow rate), and effort variables (e.g., voltage, torque or force, and fuel heating value); this description is consistent because the product of flow times effort is equal to the power in the respective domains. Each element is characterized by considering its power conversion efficiency, its performance characteristics, and possibly a normalized cost versus flow and effort domains. Especially, if we concentrate on what happens upstream of the electric machines (that are reversible energy converters from electric to mechanical energy domains), we can introduce three energy domains: electric, chemical, and thermal (heat).

In the next discussion, we will use the term fuel cell system to define the whole system including the converter (the actual fuel cell stack) and the chemical energy storage system (hydrogen, natural gas, methanol, or gasoline according to the type of fuel cell).

4.2.1 Converters

Converters are elements that perform energy conversion between different energy domains. Therefore, both fuel cells and photo-voltaic panels belong to this category. DC/DC (direct current) converters and power-link can be classified as intra-domain or transformers because the energy remains in the same domain (electric). A control action takes place in transformers that are often multi-port elements concerning the power flow. Each converter possesses an efficiency-versus-effort and/or flow characterization netlists of other properties and constraints (e.g., maximum current). Another important specification of an energy converter is its power-to-weight ratio or power density because it affects the final take-off weight of the power system. Finally, the dynamic response of the converter is also important in UAVs because of the large variability of the power demand.

The main converter addressed in this chapter is the fuel cell. A fuel cell system is a good choice for small UAVs due to its silent operation, absence of pollutant emissions, limited greenhouse gas emission (only water vapor), and high efficiency when compared to internal combustion engines.

In fuel cell systems, the fuel cell stacks are the actual devices that convert chemical energy into electrical energy through the reverse electrolysis reaction of water. Fuel cells are generally classified according to the fuel used, the operating temperature, and the type of electrolyte. The proton exchange membrane fuel cell (PEMFC) stacks are generally used in UAVs. Because it is more suitable for small-type UAVs due to its low operating temperature and high efficiency compared to other fuel cell types. The direct methanol fuel cell (DMFC) and solid oxide fuel cell (SOFC) are other fuel cell types used in UAVs. Although the efficiencies of SOFC and PMFC fuel cells are close to each other, the operating temperature of SOFC is generally 500–100°C degrees. Most of the time, these temperatures are not suitable for UAVs. Also, additional equipment for cooling will increase the mass of UAV. Studies of [48, 49] can be given as an example of the use of SOFC in UAVs. Another type of fuel cell used in UAVs is the DMFC. The operating temperatures of PEMFC and DMFC are approximately the same. Except that, the power density and efficiency of DMFC are lower than PEMFC. The power density of DMFC is about 50 W/kg, its efficiency is 20–30%, while the power density of PEMFC is about 150 W/kg, its efficiency is 40–60% [24, 50]. The reason for these differences is that methanol is used as fuel in DMFC. Although methanol is easier to store than hydrogen, its reaction rate is low. For this reason, the efficiency and power density of DMFC are lower than PEMFC. A comparison of the properties of PEMFC, DMFC and SOFC is given in Table 4.1. Studies of [51–54] can be given as an example of the use of DMFC in UAVs.

The basic structure of a PEMFC is given in Figure 4.1. Generally, the cell consists of two electrodes with catalysts, electrolytes, and flow channels. The catalyst at the anode decomposes hydrogen into proton H^+ and electrons e^-. The fuel cell electrolyte allows only protons to pass while electrons pass to the cathode with the help of an external circuit. Thus, electricity is produced by the movement of electrons from anode to cathode. Then the protons and oxygens together form water at the cathode. Water and heat are thrown into the environment as waste from the fuel cell. In this way, electricity can be produced by continuously supplying hydrogen and air to the system.

The overall chemical reaction occurring in fuel cells is given in Equation (4.1). Accordingly, from the reaction of 1 mol of hydrogen with oxygen, 1 mol of water and 286 kJ/mol of heat are released. The energy (Gibbs

Table 4.1 Comparison of fuel cell types used in UAVs [24, 55, 56].

Type	Fuel and oxidizer	Efficiency (%)	Temp. (°C)	Stack/ system power density (W/kg)	Chemical reaction
PEMFC	Hydrogen Oxygen/Air	40-60	30-100	>500 >150	$H_2 \rightarrow 2H^+ + 2e^-$ and $\frac{1}{2} O_2 + 2H^+ + 2e^- \rightarrow H_2O$ $\frac{1}{2} O_2 + H_2 \rightarrow H_2O$
DMFC	Methanol Oxygen/Air	20-30	20-90	>70 >50	$CH_3OH + H_2O \rightarrow CO_2 + 6H^+ + 6e^-$ and $3/2\ O_2 + 6H^+ + 6e^- \rightarrow 3\ H_2O$ $CH_3OH + 3/2\ O_2 \rightarrow H_2O + CO_2$
SOFC	Hydrogen/ Hydrocarbon Oxygen/Air	30-50	500-1000	>800 >100	$H_2 + O^{2-} \rightarrow H_2O + 2e^-$ and $\frac{1}{2} O_2 + 2e^- \rightarrow O^{2-}$ $\frac{1}{2} O_2 + H_2 \rightarrow H_2O$

free energy-ΔG) that can be converted by this reaction is the difference between reaction enthalpy (ΔH) and entropy (ΔS) (Equation 4.2). In general form, the electrical work done by a cell is expressed in Equations (4.3) and (4.4) [57]. The cell voltage is expressed in the most general form by Equations (4.5) and (4.6). The theoretical voltage that can be obtained from a cell is expressed by Equations (4.15) (Nernst equation) by including the effects of temperature and pressure in Equation (4.6) [57]. According to Equation (4.7), the voltage that can be obtained from a cell is 1.229 V

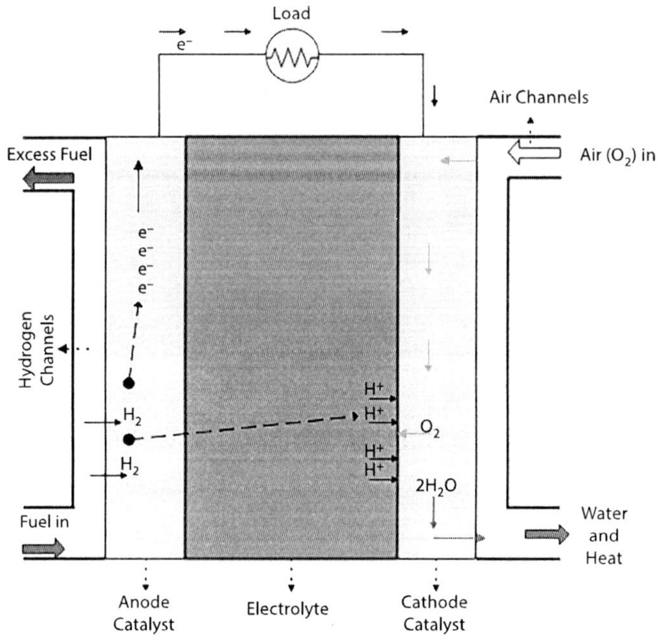

Figure 4.1 Basic structure of a fuel cell.

at standard atmospheric pressure and temperature. However, in practice, a voltage value of about 0.7 V can be obtained due to activation, ohmic, and concentration losses. By combining cells in series, stacks are created to meet the higher voltage requirement. The voltage of these stacks is the total voltage of the cells. Also, the area of cells determines the total current. The performances of the fuel cells are evaluated by the curves generated voltage (V) versus current density (mA/cm²). The cell voltage obtained by subtracting the losses is given in Equation (4.8) [57].

$$H_{2(g)} + 1/2O_{2(g)} \rightarrow H_2O_{(s)} + 286\ kJ\ mol^{-1} \tag{4.1}$$

$$\Delta G = \Delta H - T\Delta S \tag{4.2}$$

$$W_{el} = -\Delta G \tag{4.3}$$

$$W_{el} = nFE \tag{4.4}$$

$$E = \frac{-\Delta G}{nF} \tag{4.5}$$

$$E = \frac{-1}{nF}(\Delta H - T_{fc}\Delta S) \tag{4.6}$$

$$E(T,p) = \left(-\frac{\Delta H - T\Delta S}{nF}\right) + \frac{RT}{nF}\ln\left(\frac{p_{H_2}p_{O_2}^{0.5}}{p_{H_2O}}\right) \tag{4.7}$$

$$V = E - \Delta V_{act} - \Delta V_{ohmic} - \Delta V_{con} \tag{4.8}$$

where G, H, T, S, W_{el}, n, F, E, p_{H_2}, p_{H_2O}, and $p_{O_2}^{0.5}$ represent Gibbs free energy (kJ/mol), reaction enthalpy (kJ/mol), temperature (K), entropy (kJ/K mol), electric work (kJ), number of electrons per molecule of H_2, Faraday's constant (Coulombs/electron mol), cell's voltage (V), hydrogen partial pressure (Pa), water partial pressure (Pa), and oxygen partial pressure (Pa), respectively.

The fuel cell models are generally divided into three groups as chemical, experimental and electrical [58]. The chemical models are complex models that include mass transport, heat transfer, diffusion cell chemistry and thermodynamics. They are generally not suitable for real-time energy management of vehicles. The experimental models use empirical expressions and look-up tables derived from experiments. The experimental models are simpler than chemical models, and the computational burden is less. However, these models are only valid for special fuel cell stacks and in a narrow range of operation. The electrical models simulate the fuel cell using electrical components. They have a simple structure. It is the most suitable model for the simulation of real-time energy management in electric vehicles. In the literature, simulation studies on the hybrid propulsion system in UAVs generally use the model existing in MATLAB Simulink for the fuel cells [58].

One of the most critical issues in fuel cells is their slow dynamic response, therefore, differences occur between static polarization curves and dynamic performances of the fuel cells. In terms of dynamic response rate, the correct order of energy sources from large to small is a supercapacitor, battery, and fuel cell [59]. This deficiency is more important for UAVs compared to other electric vehicles, especially since UAVs have different power demands at different flight stages.

During the flight it is important to protect the fuel cell from membrane dehydration and fuel starvation [60]. Dynamic performance in fuel cells is a phenomenon dependent on water, heat, and gas management [61, 62] and is an issue that needs to be developed especially for UAVs.

4.2.2 Storage Systems

Storage systems (energy sources) include batteries and capacitors that store energy in the electric domain, and fuel tanks (like hydrogen) that contain chemical energy to be converted into electricity by an appropriate energy converter. When two or more different storage systems are used for the propulsion the power system is called a hybrid. In small UAVs, batteries are the most common energy source but many studies in the literature propose fuel cell and battery hybrid systems to improve the endurance of the UAV. Moreover, the positive effect of the supercapacitor on the dynamic response of this hybrid system has been proven with the studies (example of [27, 63]) and it is expected that the use of supercapacitors in such hybrid systems will increase. Therefore, hybrid systems consisting of fuel cell and supercapacitor, in which the supercapacitor is used as an auxiliary source, have been a popular choice recently. Although solar cells can produce energy during a flight on UAVs (airborne energy), their application areas are limited because they require much surface area and sunlight [64]. Finally, energy can be produced by placing piezo materials on the wings of the UAV. However, this energy is in the milliwatt range [65]. Therefore, they cannot be used as a main or auxiliary source in a UAV.

Fuel cell-powered UAVs have provided a significant increase in flight time and range compared to only battery-powered ones. However, the flight time and range of a fuel cell-powered UAV depend on the amount of hydrogen stored in a tank. Since hydrogen gas has a low density of 0.089 kg/m [23], the storage tank volume should be large in UAVs with long flight times. In addition, since hydrogen is stored in these tanks at high pressures such as 35–70 MPa [14], it requires a large storage tank mass to ensure strength. Hydrogen can be stored as liquid by lowering the temperature to 20 K for 1 bar.

The large mass and volume of the storage tank significantly limit the use of fuel cells in UAVs. Therefore, besides conventional compression storage, cryogenic liquid and chemical hydrides storage methods have been developed for hydrogen storage. In the cryogenic liquid hydrogen storage method, hydrogen is stored as a liquid at low temperatures. Thus, the volume of the storage tank is reduced, but it is difficult to maintain the cryogenic temperature. Moreover, additional energy is required to liquefy the hydrogen and hydrogen losses occur [66]. Using solid chemical hydrides such as $LiBH_4$, $NaBH_4$, KBH_4 and $RbBH_4$, [34], hydrogen can be stored without falling to cryogenic temperatures [55]. However, in this method, some additional equipment is required to obtain hydrogen. Detailed comparisons and specifications of conventional compressed, cryogenic liquid, and chemical hydride hydrogen storage methods can be found in tabular form in [14, 56, 67].

Storage systems are characterized by technical specifications that can be used as Key Performance Indexes (KPIs) for their comparison within this specific application:

- gravimetric energy density (or energy per unit mass)
- volumetric energy density (or energy per unit volume)
- power density (or power per unit mass/volume)
- storage efficiency (mass of the energy carrier divided by the total mass of the energy carrier and its tank)
- round trip efficiency (ratio of energy released during discharge to energy stored during charging)
- lifetime (the useful life of the energy components in years or cycles)

We can notice that these performance indexes are usually competitive. For example, supercapacitors have the highest lifetime and very high values of power density but poor energy density. Even among batteries of the same family (like for example Lithium-ion batteries), it is possible to define a trade-off between power density and energy density according to their design and manufacturing. In the case of batteries and supercapacitors, energy and power are linked so it is necessary to keep in mind the relation between them and, also include the power density in the KPIs.

The comparison of energy components in terms of gravimetric energy density, volumetric energy density, total efficiency, and lifetime is given in Figure 4.2. In this figure, adapted from literature, "fuel cell"

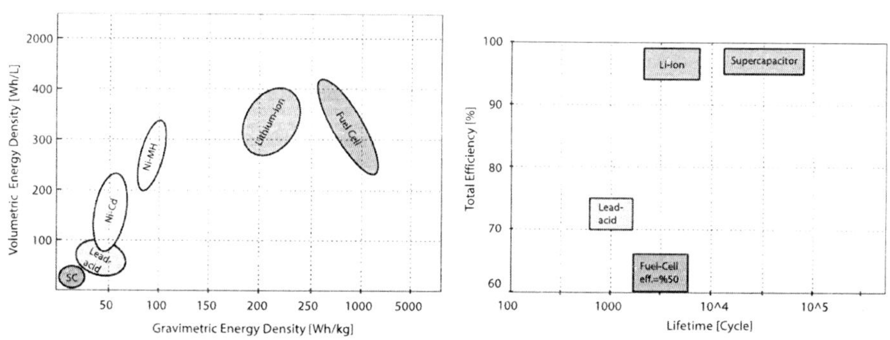

Figure 4.2 Comparison of energy components in terms of efficiency, lifetime, gravimetric energy density, and volumetric energy density [68–70].

is meant as the whole system including the converter (fuel cell) and its power source (hydrogen stored in a pressurized vessel). Table 4.2 and Table 4.3 compares the above-cited components including also solar cells and internal combustion engine where this term is used to denote, as in the case of the fuel cell, the ensemble of energy source (liquid fossil fuel) and converter.

Table 4.2 Comparison of energy components used in UAVs [20, 24, 50, 71].

Energy sources	Main advantages	Main disadvantages
Battery (B)	Medium energy density.	Low power density.
Fuel Cell system (FC)	High energy density.	Slow dynamic response times. Problems associated with production and storage of hydrogen.
Solar Cell (SC)	Free availability of the solar energy (airborne energy production).	Lack of reliability, large surface area and maximum power point tracking (MPPT) technique (compatible only with HALE UAVS).
Supercapacitor (S)	Very high-power density.	Very low energy density.
Internal combustion engine systems (ICE)	High energy and power densities.	Vibratory operation, high thermal trace, harmful emissions.

Table 4.3 Power and energy density of energy sources [71].

Energy sources	Power density (W/kg)	Energy density (Wh/kg)
Fuel Cells	10-100	600-1000
Battery	1-500	10-200
Supercapacitor	110-10000	2-8
Solar Cell	50-200 (W/m^2)	-
Internal Combustion Engine	500-900	500-2000

4.3 Fuel Cell-Based Hybrid Propulsion System Architectures

As explained before, the different energy sources described above can be combined in hybrid electric power systems. Such systems can have a variety of topologies but are generally classified according to the power and energy degree of hybridization [72]. A hybrid electric power system needs a suitable energy management unit to increase efficiency. Figure 4.3 shows how the different systems are connected to the electric motor through the energy management unit. Note that each component is controlled through DC/DC converters in the UAV propulsion system. In addition, total efficiency of each component is approximately shown in Figure 4.3.

In these hybrid systems, the fuel cell is used generally as the main source, while the battery and supercapacitor are used as the auxiliary source. The hybrid system created by fuel cells and batteries is shown in Figure 4.4. In these hybrid topologies, active energy management is implemented with the help of DC/DC converters, as explained later. In Figure 4.4-a, the battery is connected to the bus with a DC/DC converter. Thus, the battery can

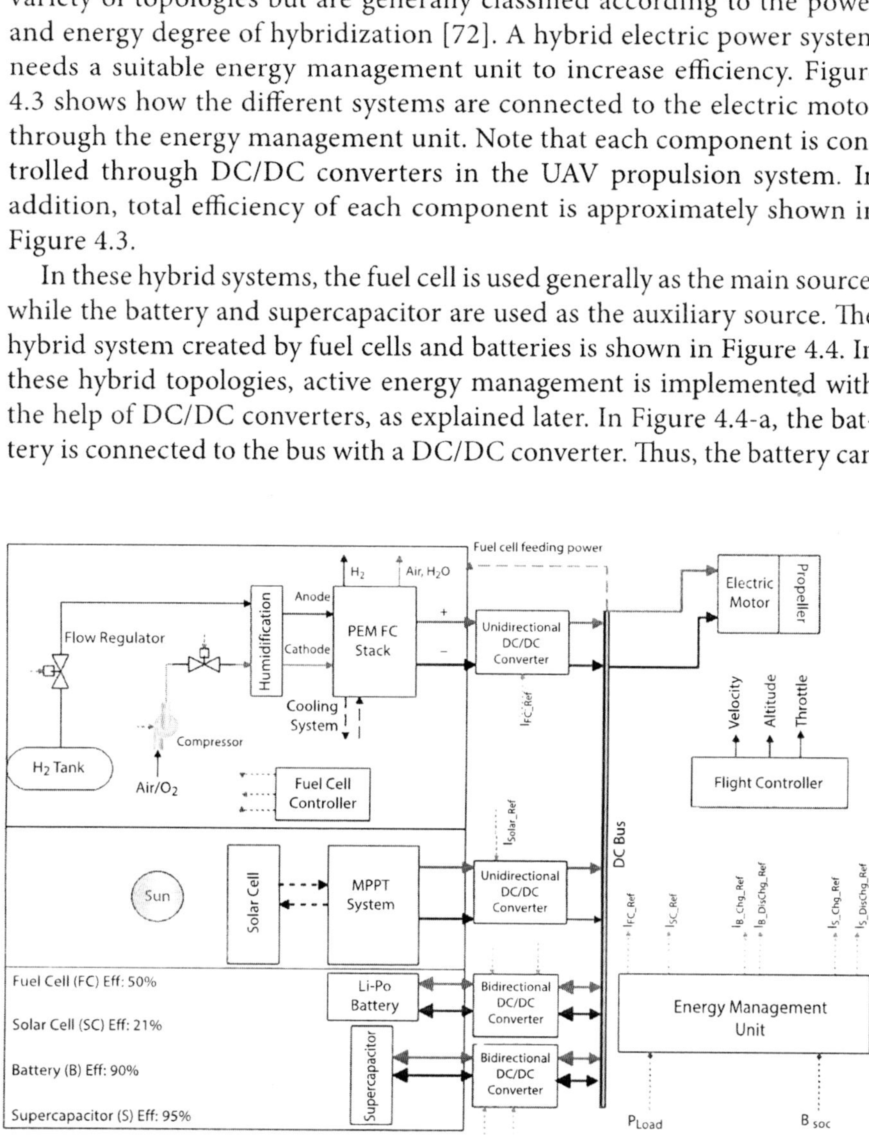

Figure 4.3 Demonstration of an energy management unit and energy components in a UAV with hybrid propulsion.

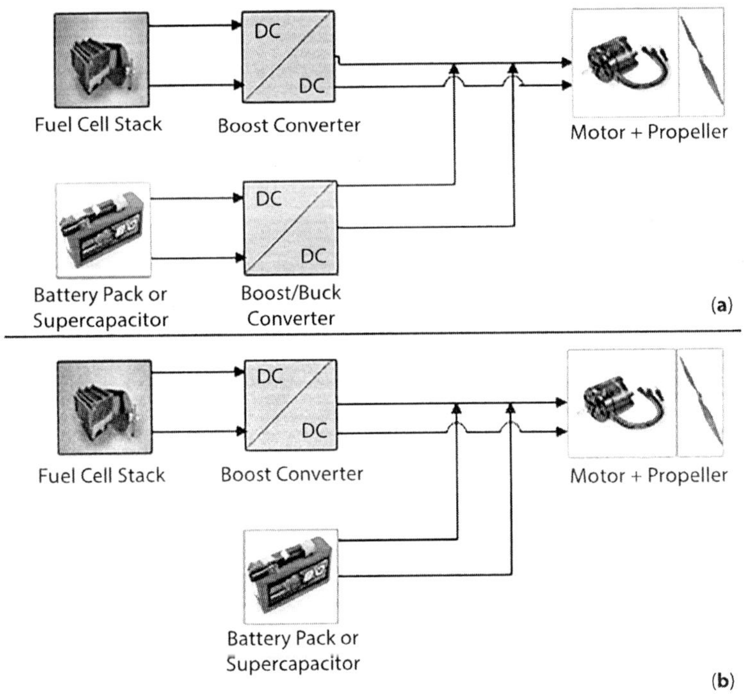

Figure 4.4 Classification of hybridization schemes according to the control of fuel cell and battery or supercapacitor: (a) only the fuel cell is controlled, (b) both the fuel cell and the battery/supercapacitor are controlled.

be charged when the energy generated by the fuel cell is higher than the UAV demand energy. In Figure 4.4-b, the battery is directly connected to the DC bus. This topology can be selected for applications where charging the battery is not critical.

The mission of the UAV should be considered in the selection of energy sources and topologies. For some missions, the demand power of the UAV can vary continuously and there may be high power requirements instantaneously. Thus, if the demand power of the UAV is dynamic, supercapacitors can be added to the hybrid system consisting of fuel cells and batteries. The triple hybrid system topologies that the fuel cell, battery, and supercapacitor constitute in UAV are shown in Figure 4.5. The power density of supercapacitors is greater than the other energy source. Therefore, by including the supercapacitor in the hybrid system consisting of fuel cell and battery, the hybrid system can still become more stable for dynamic loads. In addition, the use of supercapacitors has a positive effect on the lifetime of other resources and UAV maneuverability. In missions that require long endurance at high altitudes, solar cells are a good option. The

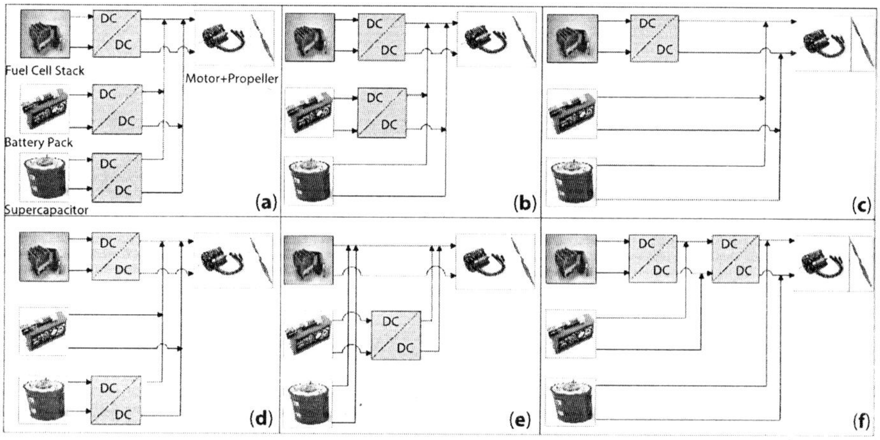

Figure 4.5 Classification of hybridization schemes according to the control of fuel cell, battery, and supercapacitor: (a) fuel cell, battery, and supercapacitor (b) fuel cell and battery (c) only fuel cell (d) fuel cell and supercapacitor (e) only battery (f) fuel cell and battery.

solar cells can be added to the hybrid systems shown in Figure 4.4 and various combinations can be created according to UAV's mission.

4.4 Experiments on Fuel Cell-Based UAVs

The first UAV with a fuel cell (Hornet named) was developed in 2003 [73]. Since then, many UAVs with fuel cell have been developed and produced successfully. Hornet had a flight time of 0.25 h. UAV named Ion Tiger which was developed in 2013 had a flight time of 48 hours. In shorts, the flight times of UAVs with fuel cell have improved significantly since 2003. However, there are still issues to be developed regarding the use of fuel cells in UAVs. Examples of these are the dynamic response speed of fuel cells, efficiency increase, storage of hydrogen, and control of the fuel cell [24]. The UAVs with fuel cells that have been produced and tested in the academic field in recent years and their characteristics are shown in Table 4.4. Considering the number of studies, it is seen that there is intense interest in fuel cell-powered UAVs in recent years.

Several studies were conducted in the development of fuel cell-powered UAV design and flight tests, especially in Georgia Institute of Technology, Colorado State University, California State University, Oklahoma State University, and The Korea Aerospace Research Institute (KARI). These studies are related to the design of the fuel cell, energy management, and its integration into the UAV. In studies where the focus is not on fuel cell design and performance improvement, fuel cell modules produced by companies such as Horizon, Lynntech, Protonex, Spectronik, Intelligent Energy, and FCair are used generally.

Table 4.4 Examination of UAVs with fuel cells [74–79].

FC power (W) FC mass (g) UAV mass (kg) Wing span (m)	Type producer	Contribution	Ref.
150 W 320 g 6.5 kg 2.77 m	PEMFC Spectronik FLY-150	Passive energy management technique was applied to the hybrid system consisting of a battery, fuel-cell, and supercapacitor. The effect of the supercapacitor was investigated in detail in this hybrid system.	[80]
250 W 730 g 6.5 kg 3 m	PEMFC Horizon Aerostak A-250	A detailed design plan was presented for a UAV with a fuel cell.	[81]
200 W - 18.5 kg 6.9 m	PEMFC Horizon Aerostak A-200	Active energy management technique was applied for the first time for small UAVs with a fully electric hybrid propulsion system.	[37]
100 W 0.9 kg 7.75 kg -	PEMFC Horizon H-100 PEM	Power and energy management were investigated by controlling the fuel cell variables.	[82]
218 W 0.6 kg - 2.5 m	PEMFC Authors	A fuel cell based on sodium borohydride was developed for a UAV and flight tests of the UAV were conducted.	[78]
500 W - - -	PEMFC HES A-500	Hardware-in-the-loop tests of the hybrid propulsion system consisting of battery and fuel cell were carried out. Rule-based control was examined in the energy management of a UAV.	[83]
100 W - - 2 m	PEMFC Horizon Aerostak A-100	Sodium borohydride ($NaBH_4$) was used as a hydrogen source. The performance of the fuel cell and battery hybrid system has been studied.	[34]

(Continued)

Table 4.4 Examination of UAVs with fuel cells [74–79]. (*Continued*)

FC power (W) **FC mass (g)** **UAV mass (kg)** **Wing span (m)**	**Type** **producer**	**Contribution**	**Ref.**
200 W 9.8 kg 2 kg (only FC) 4.7 m	PEMFC Horizon Aeropak 200	The performance of the propulsion system of a UAV, which consists of a fuel cell and battery, was examined.	[79]
245 W 2.6 kg - -	SOFC Authors	The use of SOFC for small UAVs was investigated and CFD studies of the SOFC were conducted.	[49]
100 W 470 g 2.1 kg 1.52 m	PEMFC (NaBH4 based) Horizon H-100 PEM	The performance of hydrogen generation was improved using the Co-B catalyst supported on highly porous ceramic material. Passive power management was applied to the hybrid system consisting of a battery and fuel cell.	[84]
650 W 0.8 kg 1.9 kg -	PEMFC Intelligent Energy 650 W	Using real flight power demand, a frequency separation rule-based approach was proposed to conduct energy management of a drone.	[85]
800 W	Intelligent Energy 800 W	A UAV with vertical take-off capability and fuel cell/battery hybrid power system was developed and flight tests were carried out.	[86]
300 W 11 kg 2.8 kg 4.2 m	PEMFC Authors	The hybrid system consisting of a battery, fuel cell, and solar panels was controlled according to the power switching method.	[87]
465 W (Peak Power) 4.96 kg 16.4 kg 6.58 m	PEMFC Authors	A comprehensive comparison of design methods for UAVs with fuel cells was presented.	[88]

4.5 Energy Management Strategies of Fuel Cell-Based Propulsion

Hybrid systems need a supervisory control which implements a suitable energy management strategy, i.e. a distribution of the power request among the different energy sources during the mission. Generally, energy management strategies applied in UAVs and other electric vehicles are classified into three groups: rule-based (heuristic), learning-based, and optimization-based. Classification of energy management strategies is shown in Figure 4.6.

The comparison of these strategies in terms of UAVs and application examples in UAVs are given in Table 4.5. According to Table 4.5, studies on the energy management of UAVs have usually focused on rule-based approaches. Although there are many studies on meta-heuristic based energy management in other electric vehicles (example of [89–93]), there are very few studies in which the energy management of UAVs is conducted with this method. Therefore, optimization and learning-based approaches in the energy management of UAVs have the potential to become new research topics. In this section, energy management strategies and hybrid system topologies are discussed for UAVs.

In optimization-based strategies, an objective function is minimized or maximized according to some constraints. The objective function can include variables such as endurance power consumption, fuel consumption, state of charge of battery or supercapacitor and emission value [114]. In general, optimization-based strategies are divided into two groups: global optimization (causal) and real-time optimization (non-causal). The

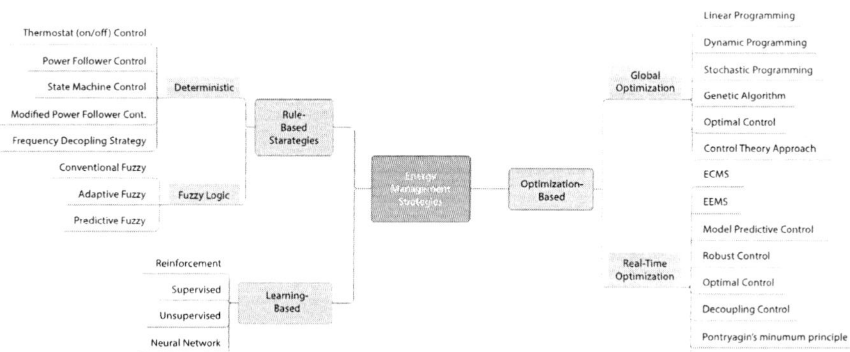

Figure 4.6 Classification of energy management strategies.

Table 4.5 Example of energy management strategies (EM) in UAVs.

EM method	Advantages	Drawbacks	Examples
Deterministic Rule-Based	Computing burden is low. Easy and simple to implement. Real-time applicability.	Cannot guarantee optimality. Requires calibration and parameter adjustment. Low overall system efficiency.	[22, 36–38, 43, 76, 87, 94–98]
Fuzz Logic Rule-Based	Real-time applicability. Com-putting burden is low. Suitable for multivariable and nonlinear systems. It has the robustness to measurement noise.	Cannot guarantee optimality. Requires membership function and rules calibration. Rules are based on human experience and prior knowledge.	[39, 99–102]
Learning-Based	Training. Adaptive capability. Model-free control. Real-time applicability.	Convergence and learning times may be high. No stability. Method complexity.	[103–105, 129]
Global Optimization-Based	Can guarantee optimality. Does not require calibration.	Computing burden is high Cannot apply real-time applicability. It is mathematically complex.	[106–110]
Real-time Optimization-Based	Can guarantee sub-optimality, real-time applicability.	Cannot guarantee global optimality.	[111–113]

global optimization strategies largely depend on the a-priori knowledge of flight profiles [115, 116]. It guarantees a global optimum, but it cannot be implemented in real-time because the computing loads are too high. The real-time optimization strategies are obtained by converting the problem in a local minimization of an instantaneous objective function that is solved in real-time and has a low computational burden. Rule-based energy management strategies are based on predefined rules that are obtained offline and then implemented online as maps or finite-state machines. Since these energy management strategies have a relatively simple structure and provide a low computational burden, they can be implemented in real-time. Rule-based energy management strategies are generally divided into two general groups as deterministic and fuzzy logic. In learning-based (L-B) energy management techniques that use large data sets to find optimum are not needed predefined rules [117]. These techniques have the advantages of learning, adaptive capability and model-free control [118]. However, it is difficult to create an accurate database with these techniques. According to learning techniques, it is classified as reinforcement-learning, supervised learning, unsupervised learning, and neural networks. In reinforcement-learning (RL) strategies, learning comes from the interaction between an agent and an environment. A learning agent and an environment are in constant interaction [117]. It does not need a pre-determined data set for learning. An agent can learn from real-time data. Therefore, RL is very suitable for real-time applications. The advantages of the

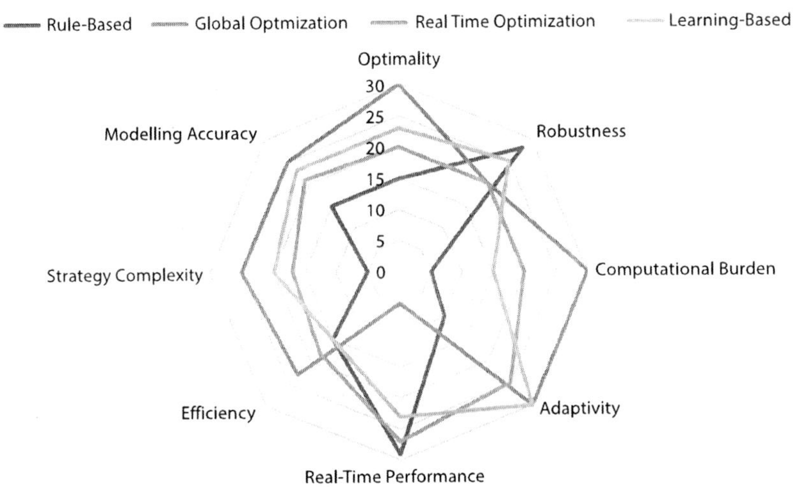

Figure 4.7 Comparison of energy management strategies according to performance criteria [118, 120].

Table 4.6 Studies on energy management (EM) in UAVs.

Ref.	System topology	EM rype	Exp./ Sim.	Contribution
[121]	FC+B	Passive Method	Both	Sodium borohydride was used for hydrogen storage.
[28]	B+FC+S	Passive Method	Sim./ HWIL	The effect of battery capacity on dynamic response of fuel cell-powered hybrid system was examined in detail and hardware-in-loop simulation was performed.
[27]	B+FC+S	Passive Method	Sim./ HWIL	The performance of the UAV with hybrid propulsion was examined in detail. In addition, the effect of the supercapacitor on the battery and fuel cell hybrid system was investigated.
[122]	FC+B+SC	Passive Method	Sim.	The flight time of a UAV was increased from 470 minutes to 970 minutes.
[123]	FC+B	Passive Method	Exp.	Flight and wind tunnel tests of a UAV, which possessed a NaBH4 hydrogen generator and battery, were performed.

(*Continued*)

Table 4.6 Studies on energy management (EM) in UAVs. (*Continued*)

Ref.	System topology	EM rype	Exp./ Sim.	Contribution
[36, 37]	FC+SC+B	Rule-based Method	Both	The active energy management technique was proven to be more efficient than the passive one.
[85]	FC+B	Rule-based Method	Both	A new energy management strategy based on frequency separation rule-based was developed for a fuel cell-powered UAV.
[39]	SC+B+FC	Fuzzy State Machine	Sim.	A new energy management strategy, a combination of the state machine and fuzzy logic, was developed and a full UAV simulation model was created for trajectory optimization and energy management of the UAV.
[124]	FC+B	Fuzzy Logic	Sim.	The fuzzy logic-based energy management technique was developed for a UAV.

(*Continued*)

Table 4.6 Studies on energy management (EM) in UAVs. (*Continued*)

Ref.	System topology	EM rype	Exp./ Sim.	Contribution
[125]	FC+B	Time Delay Control, Proportional Integral Control	Both	The time delay control and proportional-integral control techniques were compared with each other.
[126]	FC+B+S	Meta-heuristic	Sim.	Many metaheuristic algorithms were applied in a triple hybrid system consist of a battery, super capacitor, and fuel cell.
[127, 128]	B+FC+S	Meta-heuristic	Sim.	The 9 different meta-heuristic algorithms were applied and compared with each other.
[129]	B+ICE	Q-Learning	Sim.	A full simulatio n model was established for hybrid UAVs consisting of battery/ICE and by applying the Q-Learning method, reduction in the fuel consumption and emission of the UAV was achieved.
[32, 102]	FC+B	Adaptive Neuro Fuzzy Interfence System	Both	The energy management and fuel cell control were carried out together for a UAV.

Table 4.7 Some patent studies in small UAVs with the hybrid propulsion system.

Patent number and title	Contribution
KR101904225B1 The hydrogen fuel cell drone equipped with the hybrid controller	In the rotary-wing UAV, a controller has been developed for the hybrid system consisting of the fuel cell as the main energy source and battery as the auxiliary source. The hybrid controller controls the battery charge/ discharge rate and fuel cell input and output variables together.
KR102049642B1 Fuel cell powerpack for drone and state information monitoring method	To increase the flight endurance of fuel cell-powered UAVs, a controller has been developed that allows the monitoring and transfer of values such as temperature, pressure, voltage, etc., which vary according to the flight conditions of the UAV, to the computer.
CN106295195A Real-time estimation method for cruising distance of UAV (unmanned aerial vehicle) carrying hydrogen fuel cell	For UAVs driven by a fuel cell, a real-time range estimation method has been developed by using hydrogen tank pressure and average velocity.
CN107193285A Multi-rotor fuel cell plant protection UAV (unmanned aerial vehicle) control system and work method	A system consisting of a hybrid power supply, data acquisition unit, and main control unit has been developed for UAVs. This system provides altitude and position control, endurance enhancement, and optimized power-sharing between energy sources.
CN104943857A Petrol-electric hybrid five-rotor unmanned aerial vehicle	Hybridization consisting of the ICE and battery has been created for the rotary-wing UAV. Through this hybrid system, problems such as short-range, low payload capacity, and unstable in-flight attitude have been solved.

reinforcement-learning technique are real-time learning and model-free control. In supervised learning, training continues until the accuracy level of training data reaches the desired level. In unsupervised learning, on the other hand, learning is done by drawing conclusions from the input data.

The Neural network is suitable for systems with multiple inputs and multiple outputs [119]. It has an adaptive structure and can be used to control any system. As the calculation load is low, The Neural network can be applied in real-time. However, its performance depends on learning data and prediction accuracy [104]. Also, convergence and training can take over time.

Performance evaluation of energy management techniques is analyzed according to criteria such as computational load, robustness, efficiency, and optimality. The comparison according to these performance criteria is shown in Figure 4.7. When evaluated according to these performance criteria, each algorithm has its superiority and deficiency. For this reason, the UAV mission and characteristics should be considered in the selection of the energy management strategy. In addition, hybrid energy management algorithms can be created according to the mission of UAVs.

A summary of recent academic studies on energy management of UAVs with hybrid propulsion systems can be found in Table 4.6. Additionally, some patent studies on the hybrid propulsion system and design in UAVs in both the fixed-wing and rotary-wing category are given in Table 4.7. In addition, there are many studies on power-sharing and control issues in the hybrid system formed by the fuel cell with other energy sources such as batteries or supercapacitors.

4.6 Conclusions and Future Trends for Fuel Cell-Based Propulsion of UAVs

In this chapter, the energy converters/sources (batteries, supercapacitors, fuel cells, and solar cells), hybrid topologies, and energy management strategies are examined in terms of small electric unmanned aerial vehicles (UAVs).

According to the results obtained from this research, future studies on the energy management of UAVs are listed as follows.

1) The hybrid propulsion simulation models can be set up as a subsystem within the full UAV model. So that the effects of wind, throttle, and altitude can be examined on energy components and energy management strategies. The fuel cell performance can be investigated in HALE-type UAVs, especially since the effect of altitude on the fuel cell is important.

2) The hybridization of the propulsion system of small UAVs generally consists of fuel cell and battery. There are very few studies involving supercapacitors in the UAV propulsion system. In particular, when the power demand is dynamic, the use of supercapacitors has positive effects on UAV flight time and the service life of other energy sources. For these reasons, research can be conducted on the use of supercapacitors in small UAVs.
3) In choosing the energy management strategy, the mission of the UAV should be taken into account. Although an energy management technique is suitable for a UAV with a predetermined mission, it may not be suitable for a UAV with a dynamic mission. For this reason, mission-based energy management techniques should be investigated as future works in UAVs.
4) Sizing energy components and determining their capacities are important in UAVs with hybrid propulsion systems. Although there are studies on this subject in UAVs with internal combustion engines, it has been observed that there are very few in fully electric hybrid UAVs. Therefore, conceptual design studies can be conducted for hybrid fuel cell-powered UAVs.
5) Computing burden is an important criterion for the real-time applicability of energy management techniques. Therefore, studies can be conducted to reduce the computational burden in energy management strategies. In this context, hybrid energy management techniques can be developed for small UAVs.
6) It results mandatory to quantify, by experimental and numerical studies, the actual dynamic response of a fuel cell. This is important to develop suitable energy management strategies that smooth down the power request and increase the fuel cell lifetime. In addition, hydrogen storage technologies that provide less mass and volume can be examined in order to increase the flight time of fuel cell-powered UAVs.

References

1. Korchenko, A.G. and Illyash, O.S., The generalized classification of unmanned air vehicles. *2013 IEEE 2nd Int. Conf.*, pp. 28–34, 2013.
2. Babayomi, O.O. and Makarfi, A.U., Energy efficiency in unmanned aircraft systems: A review. *IEEE PES/IAS PowerAfrica Conf.*, pp. 569–574, 2019.
3. Pajares, G., Overview and current status of remote sensing applications based on unmanned aerial vehicles (UAVs). *Photogramm. Eng. Remote Sens.*, 81, 281–329, 2015.
4. Nex, F. and Remondino, F., UAV for 3D mapping applications: A review. *Appl., Geomat.*, 6, 1–15, 2014.
5. Rokhmana, C.A., The potential of UAV-based remote sensing for supporting precision agriculture in Indonesia. *Proc. Environ. Sci.*, 24, 245–253, 2015.
6. Hassanalian, M. and Abdelkefi, A., Classifications, applications, and design challenges of drones: A review. *Prog. Aerosp. Sci.*, 91, 99–131, 2017.
7. Mukherjee, A., Misra, S., Raghuwanshi, N.S., A survey of unmanned aerial sensing solutions in precision agriculture. *J. Netw. Comput. Appl.*, 148, 102461, 2019.
8. Manigandan, S., Thaloor Ramesh, P.K., Chi, N.T.L., Brindhadevi, K., Early detection of SARS-CoV-2 without human intervention to combat COVID-19 using drone technology. *Aircr. Eng. Aerosp. Technol.*, 93, 85–88, 2021.
9. Angurala, M., Bala, M., Bamber, S.S., Kaur, R., Singh, P., An internet of things assisted drone based approach to reduce rapid spread of COVID-19. *J. Saf. Sci. Resilience*, 1, 31–35, 2020.
10. Kumar, A., Sharma, K., Singh, H., Naugriya, S.G., Gill, S.S., Buyya, R., A drone-based networked system and methods for combating coronavirus disease (COVID-19) pandemic. *Future Gener. Comput. Syst.*, 115, 1–19, 2021.
11. Hwang, J. and Kim, H., The effects of expected benefits on image, desire, and behavioral intentions in the field of drone food delivery services after the outbreak of COVID-19. *Sustain.*, 13, 1–13, 2021.
12. Hwang, J., Kim, D., Kim, J.J., How to form behavioral intentions in the field of drone food delivery services: The moderating role of the COVID-19 outbreak. *Int. J. Environ. Res. Public Health*, 17, 1–15, 2020.
13. Donateo, T., De Pascalis, C.L., Ficarella, A., Synergy effects in electric and hybrid electric aircraft. *Aerospace*, 6, 32, 2019.
14. Nadir, M., Zhou, Z., Benbouzid, M., A critical review on unmanned aerial vehicles power supply and energy management: Solutions, strategies, and prospects. *Appl. Energy*, 255, 113823, 2019.
15. Wang, Q., Xiao, D., Qi, W., Research of energy management strategy for fuel cell hybrid electric vehicle. *Chin. J. Power Sources*, 36, 1459–1462, 2012.
16. Donateo, T. and Totaro, R., Hybridization of training aircraft with real world flight profiles. *Aircr. Eng. Aerosp. Technol.*, 91, 353–365, 2019.
17. Donateo, T. and Spedicato, L., Fuel economy of hybrid electric flight. *Appl. Energy*, 206, 723–738, 2017.

18. Azizi, I. and Radjeai, H., A new strategy for battery and supercapacitor energy management for an urban electric vehicle. *Electr. Eng.*, 100, 667–676, 2018.
19. Dündar, Ö., Bilici, M., Ünler, T., Design and performance analyses of a fixed wing battery VTOL UAV. *Eng. Sci. Technol. Int. J.*, 23, 1182–1193, 2020.
20. Duy, V.N. and Kim, H.M., Review on the hybrid-electric propulsion system and renewables and energy storage for unmanned aerial vehicles. *Int. J. Electrochem. Sci.*, 15, 5296–5319, 2020.
21. Bradley, T.H., Moffitt, B.A., Fuller, T.F., Mavris, D., Parekh, D.E., Design studies for hydrogen fuel cell powered unmanned aerial vehicles. *26th AIAA Applied Aerodynamics Conference*, 18-21 August 2008.
22. Boukoberine, M.N., Zhou, Z., Benbouzid, M., Donateo, T., A frequency separation rule-based power management strategy for a hybrid fuel cell-powered drone. *IECON Proc. (Industrial Electron. Conf.)*, pp. 4975–4980, 2020.
23. Gong, A. and Verstraete, D., Fuel cell propulsion in small fixed-wing unmanned aerial vehicles: Current status and research needs. *Int. J. Hydrogen Energy*, 42, 21311–21333, 2017.
24. Zhou, M. and Prasad, J.V.R., Transient characteristics of a fuel cell powered UAV propulsion system. *J. Intell. Robot. Syst.*, 74, 209–220, 2014.
25. Wang, B., Zhao, D., Li, W., Wang, Z., Huang, Y., You, Y., Becker, S., Current technologies and challenges of applying fuel cell hybrid propulsion systems in unmanned aerial vehicles. *Prog. Aerosp. Sci.*, 116, 100620, 2020.
26. Clarke, A.J., *The conceptual design of novel future UAV's incorporating advanced technology research components*. PhD Thesis, Cranfield University, UK, 2011.
27. Gong, A., Palmer, J.L., Brian, G., Harvey, J.R., Verstraete, D., Performance of a hybrid, fuel-cell-based power system during simulated small unmanned aircraft missions. *Int. J. Hydrogen Energy*, 41, 11418–11426, 2016.
28. Gong, A., Palmer, J.L., Brian, G., Harvey, J.R., Verstraete, D., Hardware-in-the-loop simulation of a fuel- cell-based UAV propulsion. *Fourth Australas. 28. Int. Congress of The Aeronautical Sci.*, vol. 7, pp. 1–7, 2014.
29. Gong, A. and Verstraete, D., Experimental characterization of a fuel cell hybrid electrical UAV propulsion system. *52nd Aerospace Sciences Meeting*, 2014.
30. Zhang, X., Liu, L., Dai, Y., Lu, T., Experimental investigation on the online fuzzy energy management of hybrid fuel cell/battery power system for UAVs. *Int. J. Hydrogen Energy*, 43, 10094–10103, 2018.
31. Karunarathne, L., Economou, J.T., Knowles, K., Power and energy management system for fuel cell unmanned aerial vehicle. *Proc. Inst. Mech. Eng. Part G J. Aerosp. Eng.*, 226, 437–454, 2012.
32. Karunarathne, L., Economou, J.T., Knowles, K., Dynamic control of fuel cell air supply system with power management. *19th Mediterranean Conference on Control and Automation*, pp. 856–861, 2011.

33. Gadalla, M. and Zafar, S., Analysis of a hydrogen fuel cell-PV power system for small UAV. *Int. J. Hydrogen Energy*, 41, 6422–6432, 2016.
34. Kim, K., Kim, T., Lee, K., Kwon, S., Fuel cell system with sodium borohydride as hydrogen source for unmanned aerial vehicles. *J. Power Sources*, 196, 9069–9075, 2011.
35. Arat, H.T. and Sürer, M.G., Experimental investigation of fuel cell usage on an air Vehicle's hybrid propulsion system. *Int. J. Hydrogen Energy*, 45, 26370–26378, 2020.
36. Lee, B., Park, P., Kim, C., Yang, S., Ahn, S., Power managements of a hybrid electric propulsion system for UAVs. *J. Mech. Sci. Technol.*, 26, 2291–2299, 2012.
37. Lee, B., Kwon, S., Park, P., Kim, K., Active power management system for an unmanned aerial vehicle powered by solar cells, a fuel cell, and batteries. *IEEE Trans. Aerosp. Electron. Syst.*, 50, 3167–3177, 2014.
38. Lee, B., Park, P., Kim, K., Kwon, S., The flight test and power simulations of an UAV powered by solar cells, a fuel cell and batteries. *J. Mech. Sci. Technol.*, 28, 399–405, 2014.
39. Zhang, X., Liu, L., Dai, Y., Fuzzy state machine energy management strategy for hybrid electric UAVs with PV/Fuel cell/battery power system. *Int. J. Aerosp. Eng.*, 2018, 1–16, 2018.
40. Gang, B.G. and Kwon, S., Design of an energy management technique for high endurance unmanned aerial vehicles powered by fuel and solar cell systems. *Int. J. Hydrogen Energy*, 43, 9787–9796, 2018.
41. Shiau, J.K., Ma, D.M., Yang, P.Y., Wang, G.F., Gong, J.H., Design of a solar power management system for an experimental UAV. *IEEE Trans. Aerosp. Electron. Syst.*, 45, 1350, 2009.
42. Oettershagen, P., Melzer, A., Mantel, T., Rudin, K., Stastny, T., Wawrzacz, B., Hinzmann, T., Alexis, K., Siegwart, R., Perpetual flight with a small solar-powered UAV: Flight results, performance analysis and model validation. *IEEE Aerosp. Conf. Proc.*, 2016.
43. Gao, X.Z., Hou, Z.X., Guo, Z., Liu, J.X., Chen, X.Q., Energy management strategy for solar-powered high-altitude long-endurance aircraft. *Energy Convers. Manage.*, 70, 20–30, 2013.
44. Lei, T., Energy management system and strategy for a fuel cell/battery hybrid power, in: *Unmanned Aerial Systems*, pp. 289–313, Elsevier Academic Press, New York, 2021.
45. Xie, Y., Savvaris, A., Antonios, T., Dan, Z., Jason, G., Review of hybrid electric powered aircraft, its conceptual design and energy management methodologies. *Chin. J. Aeronaut.*, 34, 432–450, 2020.
46. Donateo, T., Ficarella, A., Spedicato, L., Arista, A., Ferraro, M., A new approach to calculating endurance in electric flight and comparing fuel cells and batteries. *Appl. Energy*, 187, 807–819, 2017.
47. Rizzoni, G., Guzzella, L., Baumann, B.M., Unified modeling of hybrid electric vehicle drivetrains. *IEEE/ASME Trans. Mechatron.*, 4, 246–257, 1999.

48. Ji, Z., Rokni, M.M., Qin, J., Zhang, S., Dong, P., Energy and configuration management strategy for battery/fuel cell/jet engine hybrid propulsion and power systems on aircraft. *Energy Convers. Manage.*, 225, 113393, 2020.
49. Giacoppo, G., Barbera, O., Briguglio, N., Cipiti, F., Ferraro, M., Brunaccini, G., Erdle, E., Antonucci, V., Thermal study of a SOFC system integration in a fuselage of a hybrid electric mini UAV. *Int. J. Hydrogen Energy*, 42, 28022–28033, 2017.
50. Lei, T., Yang, Z., Lin, Z., Zhang, X., State of art on energy management strategy for hybrid-powered unmanned aerial vehicle. *Chin. J. Aeronaut.*, 32, 1488–1503, 2019.
51. Kang, K., Park, S., Cho, S.O., Choi, K., Ju, H., Development of lightweight 200-W direct methanol fuel cell system for unmanned aerial vehicle applications and flight demonstration. *Fuel Cells*, 14, 694–700, 2014.
52. González-Espasandín, Ó., Leo, T.J., Raso, M.A., Navarro, E., Direct methanol fuel cell (DMFC) and H_2 proton exchange membrane fuel (PEMFC/H_2) cell performance under atmospheric flight conditions of unmanned aerial vehicles. *Renewable Energy*, 130, 762–773, 2019.
53. Kang, K., Park, S., Gwak, G., Ji, H., Ju, H., Development of a lightweight 200W direct methanol fuel cell stack for UAV applications and study of its operating characteristics (II). *Trans. Korean Hydrog. New Energy Soc.*, 23, 243–249, 2012.
54. Lee, S., Kim, D., Ro, J., Cho, Y., Kim, D., Ju, H., Development of lightweight direct methanol fuel cell (DMFC) stack using metallic bipolar plates for Unmanned Aerial Vehicles (UAVs). *Trans. Korean Hydrog. New Energy Soc.*, 28, 492–501, 2017.
55. Pan, Z.F., An, L., Wen, C.Y., Recent advances in fuel cells based propulsion systems for unmanned aerial vehicles. *Appl. Energy*, 240, 473–485, 2019.
56. Wang, B., Zhao, D., Li, W., Wang, Z., Huang, Y., You, Y., Becker, S., Current technologies and challenges of applying fuel cell hybrid propulsion systems in unmanned aerial vehicles. *Prog. Aerosp. Sci.*, 116, 100620, 2020.
57. Barbir, F., *PEM Fuel Cells Theory and Practice*. Elsevier, New York, 2013.
58. Motapon, N. M., Tremblay, O., Dessaint, L.-A., A generic fuel cell model for the simulation of fuel cell vehicles, in: *2009 IEEE Vehicle Power and Propulsion Conference*, pp. 1722–1729, 2009.
59. Thounthong, P., Raël, S., Davat, B., Energy management of fuel cell/battery/supercapacitor hybrid power source for vehicle application. *J. Power Sources*, 193, 376–85, 2009.
60. Verstraete, D., Lehmkuehler, K., Gong, A., Harvey, J.R., Brian, G., Palmer, J.L., Characterisation of a hybrid, fuel-cell-based propulsion system for small unmanned aircraft. *J. Power Sources*, 250, 204–211, 2014.
61. Tang, Y., Yuan, W., Pan, M., Wan, Z., Experimental investigation on the dynamic performance of a hybrid PEM fuel cell/battery system for lightweight electric vehicle application. *Appl. Energy*, 88, 68–76, 2011.

62. Tang, Y., Yuan, W., Pan, M., Li, Z., Chen, G., Li, Y., Experimental investigation of dynamic performance and transient responses of a kW-class PEM fuel cell stack under various load changes. *Appl. Energy*, 87, 1410–1417, 2010.
63. Khan, M.A., Khan, A., Ahmad, M., Saleem, S., Aziz, M.S., Hussain, S., Khan, F.M., A study on flight time enhancement of unmanned aerial vehicles (UAVs) using supercapacitor-based hybrid electric propulsion system (HEPS). *Arab. J. Sci. Eng.*, 46, 1179–1198, 2020.
64. Abbe, G. and Smith, H., Technological development trends in solar-powered aircraft systems. *Renewable Sustainable Energy Rev.*, 60, 770–783, 2016.
65. Li, D., Wu, Y., Da Ronch, A., Xiang, J., Energy harvesting by means of flow-induced vibrations on aerospace vehicles. *Prog. Aerosp. Sci.*, 86, 28–62, 2016.
66. Bradley, T.H., Moffitt, B.A., Mavris, D.N., Parekh, D.E., Development and experimental characterization of a fuel cell powered aircraft. *J. Power Sources*, 171, 793–801, 2007.
67. Baroutaji, A., Wilberforce, T., Ramadan, M., Olabi, A., Comprehensive investigation on hydrogen and fuel cell technology in the aviation and aerospace sectors. *Renewable Sustainable Energy Rev.*, 106, 31–40, 2019.
68. Ouyang, L., Huang, J., Wang, H., Liu, J., Zhu, M., Progress of hydrogen storage alloys for Ni-MH rechargeable power batteries in electric vehicles: A review. *Mater. Chem. Phys.*, 200, 164–178, 2017.
69. Hall, P.J. and Bain, E.J., Energy-storage technologies and electricity generation. *Energy Policy*, 36, 4352–4355, 2008.
70. Bocklisch, T., Hybrid energy storage systems for renewable energy applications. *Energy Proc.*, 73, 103–111, 2015.
71. Gao, X.Z., Hou, Z.X., Guo, Z., Chen, X.Q., Reviews of methods to extract and store energy for solar-powered aircraft. *Renewable Sustainable Energy Rev.*, 44, 96–108, 2015.
72. Pornet, C. and Isikveren, A.T., Conceptual design of hybrid-electric transport aircraft. *Prog. Aerosp. Sci.*, 79, 114–135, 2015.
73. Gundlach, J., *Designing Unmanned Aircraft Systems: A Comprehensive Approach*. American Institute of Aeronautics and Astronautics, Inc., USA, 2011.
74. Gong, A., MacNeill, R., Verstraete, D., Palmer, J.L., Analysis of a fuel-cell/battery/supercapacitor hybrid propulsion system for a UAV using a hardware-in-the-loop flight simulator. *2018 AIAA/IEEE Electr. Aircr. Technol. Symp.*, pp. 1–17, 2018.
75. Özbek, E., Yalin, G., Ekici, S., Karakoc, T.H., Evaluation of design methodology, limitations, and iterations of a hydrogen fuelled hybrid fuel cell mini UAV. *Energy*, 213, 118757, 2020.
76. Ozbek, E., Yalin, G., Umut, M., Ekici, S., Architecture design and performance analysis of a hybrid hydrogen fuel cell system for unmanned aerial vehicle. *Int. J. Hydrogen Energy*, 46, 16453–16464, 2021.
77. Kim, T. and Kwon, S., Design and development of a fuel cell-powered small unmanned aircraft. *Int. J. Hydrogen Energy*, 37, 615–622, 2012.

78. Okumus, E., Boyaci San, F.G., Okur, O., Erdor Turk, B., Cengelci, E., Kilic, M., Karadag, C., Cavdar, M., Turkmen, A., Yazici, M.S., Development of boron-based hydrogen and fuel cell system for small unmanned aerial vehicle. *Int. J. Hydrogen Energy*, 42, 2691–2697, 2017.
79. Lapeña-Rey, N., Blanco, J.A., Ferreyra, E., Lemus, J.L., Pereira, S., Serrot, E., A fuel cell powered unmanned aerial vehicle for low altitude surveillance missions. *Int. J. Hydrogen Energy*, 42, 6926–6940, 2017.
80. Gong, A.T.Y., Impact of supercapacitors on a fuel-cell-based triple hybrid for small unmanned aircraft. PhD Thesis, The University of Sydney, Australia, 2018, https://tureng.com/tr/turkce-ingilizce/australia.
81. Ozbek, E., Yalin, G., Karaoglan, M.U., Ekici, S., Colpan, O., Karakoc, T.H., Architecture design and performance analysis of a hybrid hydrogen fuel cell system for unmanned aerial vehicle. *Int. J. Hydrogen Energy*, 46, 16453–16464, 2021.
82. Karunarathne, L., Economou, J.T., Knowles, K., Intelligent power management (IPM) for transient recognition and control of PEM fuel cell/battery hybrid system. *2009 IEEE Vehicle Power and Propulsion Conference*, pp. 992–997, 2009.
83. Xie, Y., Savvaris, A., Tsourdos, A., Laycock, J., Farmer, A., Modelling and control of a hybrid electric propulsion system for unmanned aerial vehicles. *IEEE Aerosp. Conf. Proc.*, 2018.
84. Kim, T., NaBH4 (sodium borohydride) hydrogen generator with a volume-exchange fuel tank for small unmanned aerial vehicles powered by a PEM (proton exchange membrane) fuel cell. *Energy*, 69, 721–727, 2014.
85. Boukoberine, M.N., Zia, M.F., Benbouzid, M., Zhou, Z., Donateo, T., Hybrid fuel cell powered drones energy management strategy improvement and hydrogen saving using real flight test data. *Energy Convers. Manage.*, 236, 113987, 2021.
86. De Wagter, C., Remes, B., Smeur, E., van Tienen, F., Ruijsink, R., van Hecke, K., van der Horst, E., The NederDrone: A hybrid lift, hybrid energy hydrogen UAV. *Int. J. Hydrogen Energy*, 46, 16003–16018, 2021.
87. Gang, B.G. and Kwon, S., Design of an energy management technique for high endurance unmanned aerial vehicles powered by fuel and solar cell systems. *Int. J. Hydrogen Energy*, 43, 9787–9796, 2018.
88. Bradley, T.H., Moffitt, B.A., Mavris, D.N., Fuller, T.F., Parekh, D.E., Hardware-in-the-loop testing of a fuel cell aircraft powerplant. *J. Propuls. Power*, 25, 1336–1344, 2009.
89. Mohamed, A.A.A., Mohamed, Y.S., El-Gaafary, A.A.M., Hemeida, A.M., Optimal power flow using moth swarm algorithm. *Electr. Power Syst. Res.*, 142, 190–206, 2017.
90. Sousa, T., Vale, Z., Carvalho, J.P., Pinto, T., Morais, H., A hybrid simulated annealing approach to handle energy resource management considering an intensive use of electric vehicles. *Energy*, 67, 81–96, 2014.

91. Chen, S.Y., Wu, C.H., Hung, Y.H., Chung, C.T., Optimal strategies of energy management integrated with transmission control for a hybrid electric vehicle using dynamic particle swarm optimization. *Energy*, 160, 154–170, 2018.
92. Chen, Z., Xiong, R., Cao, J., Particle swarm optimization-based optimal power management of plug-in hybrid electric vehicles considering uncertain driving conditions. *Energy*, 96, 197–208, 2016.
93. Caux, S., Hankache, W., Fadel, M., Hissel, D., On-line fuzzy energy management for hybrid fuel cell systems. *Int. J. Hydrog. Energy*, 35, 2134–2143, 2010.
94. Boukoberine, M.N., Zhou, Z., Benbouzid, M., Donateo, T., Frequency separation-based power management strategy for a fuel cell-powered drone. *2020 2nd Int. Conf. Smart Power Internet Energy Syst. SPIES*, pp. 209–214, 2020.
95. Yang, Z., Lei, T., Lin, Z., Fu, H., Zhang, X., The testing platform of hybrid electric power system for a fuel cell unmanned aerial vehicle. *2018 IEEE Int. Conf. Electr. Syst. Aircraft, Railw. Sh. Propuls. Road Veh. Int. Transp. Electrif. Conf. ESARS-ITEC*, pp. 1–8, 2019.
96. Bataller-Planes, E., Lapena-Rey, N., Mosquera, J., Orti, F., Oliver, J.A., Garcia, O., Moreno, F., Portilla, J., Torroja, J., Vasic, M., Huerta, S.C., Trocki, M., Zumel, P., Cobos, J.A., Power balance of a hybrid power source in a power plant for a small propulsion aircraft. *IEEE Power Electronics Specialists Conference*, pp. 2856–2866, 2009.
97. Zhang, Y., Peng, G.O.H., Banda, J.K., Dasgupta, S., Husband, M., Su, R., Wen, C., An energy efficient power management solution for a fault-tolerant more electric engine/aircraft. *IEEE Trans. Ind. Electron.*, 66, 5663–5675, 2019.
98. Savvaris, A., Xie, Y., Malandrakis, K., Lopez, M., Tsourdos, A., Development of a fuel cell hybrid-powered unmanned aerial vehicle. *24th Mediterranean Conference on Control and Automation, MED*, pp. 1242–1247, 2016.
99. Zhang, X., Liu, L., Dai, Y., Lu, T., Experimental investigation on the online fuzzy energy management of hybrid fuel cell/battery power system for UAVs. *Int. J. Hydrogen Energy*, 43, 10094–10103, 2018.
100. Xie, Y., Savvaris, A., Tsourdos, A., Fuzzy logic based equivalent consumption optimization of a hybrid electric propulsion system for unmanned aerial vehicles. *Aerosp. Sci. Technol.*, 85, 13–23, 2019.
101. Caux, S., Hankache, W., Fadel, M., Hissel, D., On-line fuzzy energy management for hybrid fuel cell systems. *Int. J. Hydrogen Energy*, 35, 2134–2143, 2010.
102. Karunarathne, L., Economou, J.T., Knowles, K., Fuzzy logic control strategy for fuel cell/battery aerospace propulsion system. *IEEE Vehicle Power and Propulsion Conference*, 2008.
103. Harmon, F.G., Frank, A.A., Joshi, S.S., The control of a parallel hybrid-electric propulsion system for a small unmanned aerial vehicle using a CMAC neural network. *Neural Networks*, 18, 772–780, 2005.
104. Harmon, F.G., Frank, A.A., Chattot, J.J., Conceptual design and simulation of a small hybrid-electric unmanned aerial vehicle. *J. Aircr.*, 43, 1490–1498, 2006.

105. Kim, M.J. and Peng, H., Power management and design optimization of fuel cell battery hybrid vehicles. *J. Power Sources*, 165, 819–832, 2007.
106. Han, K., Qian, H., Zhang, Q., Liu, L., Hu, X., Optimization of energy management system for fuel-cell/battery hybrid power in unmanned aerial vehicle. *22nd International Conference on Electrical Machines and Systems (ICEMS)*, 2019.
107. Fang, S. and Xu, Y., Multiobjective coordinated scheduling of energy and flight for hybrid electric unmanned aircraft microgrids. *IEEE Trans. Ind. Electron.*, 66, 5686–5695, 2019.
108. Bongermino, E., Mastrorocco, F., Tomaselli, M., Monopoli, V.G., Naso, D., Model and energy management system for a parallel hybrid electric unmanned aerial vehicle. *IEEE Int. Symp. Ind. Electron.*, pp. 1868–1873, 2017.
109. Amorosi, L., Chiaraviglio, L., Galan-Jimenez, J., Optimal energy management of UAV-based cellular networks powered by solar panels and batteries: Formulation and solutions. *IEEE Access*, 7, 53698–53717, 2019.
110. Donateo, T. and Ficarella, A., A modeling approach for the effect of battery aging on the performance of a hybrid electric rotorcraft for urban air-mobility. *Aerospace*, 7, 5, 56, 2020.
111. Çınar, H. and Kandemir, I., Active energy management based on meta-heuristic algorithms of fuel cell/battery/supercapacitor energy storage system for aircraft. *Aerospace*, 8, 85, 2021.
112. Nadir, M., Fahad, M., Benbouzid, M., Zhou, Z., Donateo, T., Hybrid fuel cell powered drones energy management strategy improvement and hydrogen saving using real flight test data. *Energy Convers. Manage.*, 236, 113987, 2021.
113. Settele, F. and Bittner, M., Energy-optimal guidance of a battery-electrically driven airplane. *CEAS Aeronaut. J.*, 11, 1, 111–124, 2020.
114. Huang, Y., Wang, H., Khajepour, A., Li, B., Ji, J., Zhao, K., Hu, C., A review of power management strategies and component sizing methods for hybrid vehicles. *Renew. Sustain. Energy Rev.*, 96, 132–144, 2018.
115. Li, H., Ravey, A., N'Diaye, A., Djerdir, A., A review of energy management strategy for fuel cell hybrid electric vehicle. *IEEE Vehicle Power and Propulsion Conference (VPPC)*, 2018.
116. Zhang, P., Yan, F., Du, C., A comprehensive analysis of energy management strategies for hybrid electric vehicles based on bibliometrics. *Renew. Sustain. Energy Rev.*, 48, 88–104, 2015.
117. Tran, D.D., Vafaeipour, M., El Baghdadi, M., Barrero, R., Van Mierlo, J., Hegazy, O., Thorough state-of-the-art analysis of electric and hybrid vehicle powertrains: Topologies and integrated energy management strategies. *Renew. Sustain. Energy Rev.*, 119, 109596, 2019.
118. Sorlei, I.-S., Bizon, N., Thounthong, P., Varlam, M., Carcadea, E., Culcer, M., IIliescu, M., Raceanu, M., Fuel cell electric vehicles—A brief review of current topologies and energy management strategies. *Energies*, 14, 252, 2021.

119. Hatti, M., Meharrar, A., Tioursi, M., Power management strategy in the alternative energy photovoltaic/PEM Fuel Cell hybrid system. *Renew. Sustain. Energy Rev.*, 15, 5104–5110, 2011.
120. Sellali, M., Betka, A., Djerdir, A., Power management improvement of hybrid energy storage system based on H ∞ control. *Math. Comput. Simul.*, 167, 478–494, 2020.
121. Lapeña-Rey, N., Blanco, J.A., Ferreyra, E., Lemus, J.L., Pereira, S., Serrot, E., A fuel cell powered unmanned aerial vehicle for low altitude surveillance missions. *Int. J. Hydrogen Energy*, 42, 6926–6940, 2017.
122. Gadalla, M. and Zafar, S., Analysis of a hydrogen fuel cell-PV power system for small UAV. *Int. J. Hydrogen Energy*, 41, 6422–6432, 2016.
123. Kim, T. and Kwon, S., Design and development of a fuel cell-powered small unmanned aircraft. *Int. J. Hydrogen Energy*, 37, 615–622, 2012.
124. Corcau, J.I., Dinca, L., Grigorie, T.L., Tudosie, A.N., Fuzzy energy management for hybrid fuel cell/battery systems for more electric aircraft. *AIP Conf. Proc.*, 1836, 020056, 2017.
125. Wang, Y.X., Ou, K., Kim, Y.B., Modeling and experimental validation of hybrid proton exchange membrane fuel cell/battery system for power management control. *Int. J. Hydrogen Energy*, 40, 11713–11721, 2015.
126. Zhao, J., Ramadan, H. S., Becherif, M., Metaheuristic-based energy management strategies for fuel cell emergency power unit in electrical aircraft. *Int. J. Hydrogen Energy*, 44, 4, 2390–2406, 2019.
127. Rezk, H., Nassef, A.M., Abdelkareem, M.A., Alami, A.H., Fathy, A., Comparison among various energy management strategies for reducing hydrogen consumption in a hybrid fuel cell/supercapacitor/battery system. *Int. J. Hydrogen Energy*, 46, 6110–6126, 2019.
128. Nassef, A.M., Fathy, A., Rezk, H., An effective energy management strategy based on mine-blast optimization technique applied to hybrid PEMFC/Supercapacitor/Batteries System. *Energies*, 12, 3796, 2019.
129. Bai, M., Yang, W., Song, D., Kosuda, M., Szabo, S., Lipovsky, P., Kasaei, A., Research on energy management of hybrid unmanned aerial vehicles to improve energy-saving and emission reduction performance. *Int. J. Environ. Res. Public Health*, 17, 8, 2917, 2020.

5

Test and Evaluation of Hydrogen Fuel Cell Vehicles

Dong Hao, Yanyi Zhang, Renguang Wang*, Tian Sun and Minghui Ma

China Automotive Technology and Research Center Co., Ltd. Tianjin, China

Abstract

Firstly, the hydrogen fuel cell vehicles (FCVs) are automotives, secondly they are one type of electric vehicles, and at last they are the FCVs, and their features of high pressure of hydrogen and subzero clod start make the performance test of FCVs real different. Because of hydrogen fuel in the fuel cell vehicles with water produced, the safety, energy consumption and subzero cold start performances have become special and important requirements for the test and evaluation of the FCVs. The safety performance is mainly connected with electric safety and hydrogen safety. The energy consumption become more complicated because of the combination of two energy sources with traction battery (REESS) and hydrogen fuel cell used in the FCVs. And the water produced by fuel cell may get frozen under subzero temperature, which makes the cold start performance become more important. Therefore, the hydrogen safety in confined space, energy consumption and range, subzero cold start were selected and discussed in detail with description of test methods and application conditions.

Keywords: Hydrogen fuel cell vehicles, hydrogen safety, safety requirements, range and energy consumption, subzero cold start, test and evaluation methods

5.1 Introduction

With rapid development of hydrogen fuel cell vehicles in recent years, the shortage of standards and methods for test and evaluation of the FCVs and related main components have become more urgent. As a type of electric

---

**Corresponding author*: wangrenguang@catarc.ac.cn

vehicles, the FCV is one of the most complicated motor vehicles. And the FCVs should meet both requirements of conventional vehicles and electric vehicles, especially those for high electric voltage and motors. But the technology configurations of FCVs are different from those of the plugin hybrid electric vehicles (PHEVs) and battery electric vehicles (BEVs), which make the FCVs need some special test methods, test devices and evaluation methods (as shown in Figure 5.1).

About the requirements for general motor vehicles and the electric vehicles, many standards and technical regulations have been established by different organizations such as ISO and SAE, which are not described here. For the FCVs, the hydrogen is used to generate electric power from chemical reaction of hydrogen and oxygen, which make their safety concern become more important than PHEVs and BEVs. About the safety test requirements for the FCVs, there are four aspects as follows:

1) The first is the hydrogen safety which mainly includes hydrogen leakage in confined space, alarm function for hydrogen, installation strength of hydrogen storage system;
2) The second is about the hydrogen emission which mainly includes emission at idle condition and emission during drive cycles;
3) The third is the electric safety which mainly includes the general electric safety and wading safety;

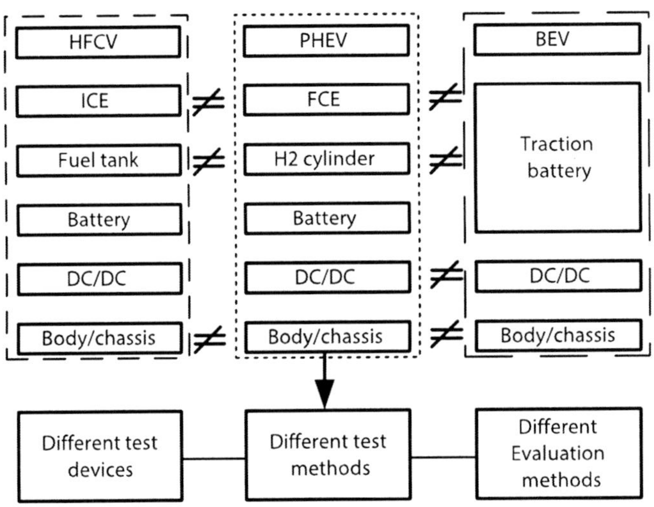

Figure 5.1 Difference of FCVs, PHEVs and BEVs (reprinted with permission [1]).

4) The fourth is the collision safety which includes passenger car collision with front, rear and side directions and commercial vehicle rollover safety.

In addition to the hydrogen safety requirements, the FCVs have special test and evaluation needs in the test of range, energy consumption, and subzero cold starting. Taking all these into consideration, the test and evaluation systems for the FCVs and its main components were established to guide the related works. With brief introduction of test system for the FCVs, fuel cell engine (FCE) and its main parts, the test methods for these three main performance of the FCVs were presented in this chapter, which includes hydrogen safety of FCVs in confined space, range and energy consumption of the FCVs, and subzero cold starting of the FCVs.

5.2 Test and Evaluation System

5.2.1 Test and Evaluation System for FCVs

The use of hydrogen as fuel for FCVs make the vehicle performance test such as the safety, range, energy consumption, environmental adaptability, dynamics, and electromagnetic compatibility (EMC) different from those of conventional vehicles and other kinds of electric vehicles. Especially the emission or leakage of hydrogen during parking and collision makes hydrogen safety more concerned; the hydrogen consumption, electric energy consumption and range are connected with the economy performance; and subzero temperature makes the environmental adaptability a problem in subzero areas. All these make it necessary that the test and evaluation system should be set up for the FCVs, fuel cell engines (FCEs), and their accessory components.

Basing on discussion and analysis, the test items were defined and corresponding test methods were listed, which form the frame of test and evaluation system for the FCVs, as shown in Table 5.1. About the test and evaluation system for FCVs, the detail can be found in reference [1], and the following section is based on this paper with some modifications.

5.2.2 Test and Evaluation System for FCEs

The FCEs are the main power source of the FCVs, their main performance includes start performance, steady performance, dynamic response performance, gas tightness, and insulation resistance etc. And the performance of

Table 5.1 Main items of test and evaluation system for FCVs.

Safety	Range & HYDROGEN consumption	Environmental adaptability	EMC	Dynamics	NVH	Road test
HYDROGEN leakage	room temp.	subzero cold tart	EMS	maximum speed	outside noise	hydrogen safety
HYDROGEN emission	subzero temp.	high temperature	EMI	acceleration	inside noise	electric safety
electric safety	high temp.	high altitude	static electricity	grade ability	stationary noise	reliability
collision safety	constant speed	air quality	human protection	top cruise speed	-	durability
refuel safety	drive cycle	HYDROGEN quality	-	-	-	economy

Table 5.2 Main test and evaluation items for the FCEs.

Start performance	Power features	Dynamic response	General performance	Environmental adaptability
cold start at ambient temperature	rated power	load increase from idle to rate power	steady performance	high temperature
hot start at ambient temperature	peak power	load increase from start to idle	efficiency	subzero temperature
cold start at subzero temperature	gravity power density	load decrease from rate power to idle	electric insulation	high altitude
hot start at subzero temperature	volume power density	load decrease from rate power to shut down	gas tightness	high humidity

FCEs (or fuel cell system) is directly connected with vehicle performance. The main items for the performance test and evaluation system of the FCEs is listed in Table 5.2.

5.2.3 Test and Evaluation System for Main Components

The main components of fuel cell engine such as fuel cell stack, air compressor, hydrogen circulator, and etc. In fact, there are many parts should be taken into this test and evaluation system, only some main parts of air and hydrogen supply were listed in this paper, as shown in Table 5.3.

5.3 Safety Performance Requirements for FCVs

About the safety requirements of the FCVs, there are some standards such as SAE J2578 <Recommended practice for general fuel cell vehicle safety> [2], ISO 23273 <Fuel cell road vehicle-safety specifications-protection against hydrogen hazards for vehicles fueled with compressed hydrogen> [3] which describe the general requirements in detail. And the China

Table 5.3 Main test and evaluation items for accessory components.

Air compressor	Hydrogen circulator	Humidifier	Radiator	Air filter	Deionizer
general performance	general performance	humidifying capacity	leakage	leakage	general performance
efficiency	efficiency	efficiency	heat dissipation	efficiency	temperature
environmental adaptability	environmental adaptability	noise	line resistance	air resistance	pressure
durability	durability	durability	durability	durability	durability
NVH	NVH	-	NVH	noise	-

national standard GB/T 24594-2020 <Fuel cell electric vehicles safety requirements>describes the main safety requirements for whole vehicle and onboard hydrogen system. The following prats of Safety performance requirements for the FCVs are based on the description of GB/T 24594-2020 [4].

5.3.1 Safety Requirements for Whole Vehicle of FCVs

The safety requirements for the FCVs include four main aspects which are hydrogen emission, hydrogen leakage, collision safety, and electric safety.

5.3.1.1 Requirements for Vehicle Hydrogen Emission

About measurement of the hydrogen concentration of emission from the tailpipe in normal operation, there are two important requirements for the test process. Firstly, the volume concentration of hydrogen emission should be measured from the start to the complete shutdown of the fuel cell engine; secondly, the sampling point should be located at 100 mm away to the tailpipe end along the extension of its geometric center line. About the test results, the requirements are that the average hydrogen volume concentration should not exceed 4% in any continuous 3s and that the instantaneous concentration should not exceed 8% during the test.

5.3.1.2 Requirements for Vehicle Hydrogen Leakage

The hydrogen leakage can occur both inside and outside the vehicle. And the requirements for hydrogen leakage to the outside of vehicles is mainly for the FCVs parked in the confined space without mechanical ventilation and with the air exchange rate per hour (ACH) less than 0.03, which is that the hydrogen volume concentration caused by hydrogen leakage should not exceed 1% at any time. The requirements for the hydrogen leakage in the vehicle are as followings:

1) The hydrogen leaked or permeated from the hydrogen system should not be directly discharged into the passenger compartment, luggage/cargo compartment, or any enclosed or semi-enclosed spaces of the vehicle with potential fire risk.
2) At least one hydrogen sensor should be installed in the top of the enclosed or semi-enclosed space where the hydrogen storage system is located.

3) A hydrogen alarm device should be installed in the easily visible area for the driver in the instrument panel.
4) It requires to alarm when the hydrogen volume concentration reaches or exceeds 2.0% ± 1.0% in the enclosed or semi-enclosed space of the FCVs.
5) The hydrogen supply system should be automatically shut off immediately when the hydrogen volume concentration reaches or exceeds 3.0% ± 1.0% in the closed or semi-enclosed space; if there are several hydrogen storage tanks, it is allowed to only shut off the one with hydrogen leakage reaching setting level.
6) The hydrogen sensor should be able to send a fault warning signal to the driver in the instrument panel when it breaks down.

5.3.1.3 *Requirements for Reminder of Low Residual Hydrogen Gas in the Tank*

The device indicating the pressure of hydrogen storage tank or the state of charge (SOC) of hydrogen storage tank should be installed in obviously visible area in the instrument panel and should be able to alert the driver via clear signal when the pressure or SOC of hydrogen storage tank reaches setting lower value.

5.3.1.4 *Requirements for Electrical Safety*

The safety requirements for electrical system of the FCVs are similar to those for battery electric vehicles (BEVs) and hybrid electric vehicles (HEVs). In China, there are three national standards describing the electrical safety which are GB 18384-2020 <electric vehicles safety requirements> [5], GB 38032-2020<electric buses safety requirements> [6] and GB 38031-2020<electric vehicles traction battery safety requirements> [7], and the electrical safety of FCVs should meet these standards according to their own conditions basing on the vehicle type.

5.3.2 Safety Requirements for Hydrogen System Safety

The requirements for hydrogen system safety are composed of six aspects, which are the installation and protection of the hydrogen storage tanks and pipelines, the pressure relief system, the hydrogen refueling and hydrogen receptacle, hydrogen leakage and detection of the hydrogen pipeline, the function of hydrogen leakage alarm device, and high pressure hydrogen discharge from storage tank.

5.3.2.1 *Requirements for the Hydrogen Storage Tanks and Pipelines*

These requirements include the three aspects:

1) As for the installation location, the pipe joints should not be installed in the completely enclosed space; the hydrogen storage tanks and pipelines should not be located in the passenger compartment, luggage compartment or other poorly ventilated space; if it is installed in a poorly ventilated place, necessary measures must be taken to discharge the possibly leaked hydrogen timely; and the hydrogen storage tanks should avoid direct exposure to the sunlight.
2) The hydrogen storage tanks and pipelines that may be affected by the heat sources such as exhaust pipe and muffler should be protected with thermal insulation.
3) High pressure pipeline and components (including hydrogen receptacle) should be grounded soundly.

5.3.2.2 *Requirements for Pressure Relief System*

The requirements for the pressure relief system include two aspects. Firstly, in order to prevent the outlets from being blocked by foreign objects, the protective measure should be taken at the outlet of release pipeline of the thermally-activated pressure relief device (TPRD) and pressure relief device (PRD). Secondly, for the gas flow direction of released hydrogen, the hydrogen released from the TPRD or PRD should not directed into any enclosed or semi-enclosed space; into or toward any of wheel covers; toward the hydrogen storage tank; toward the forward direction of vehicle; and toward the emergency exit of vehicle. The hydrogen released from the TPRD or PRD should not flow toward exposed live terminals, electric switches or other potential ignition sources; into any enclosed or semi-enclosed spaces; into or toward any of wheel covers; toward the hydrogen storage tank; and toward the emergency exit of vehicles (if any).

5.3.2.3 *Requirements for Hydrogen Refueling and Receptacle*

During hydrogen refueling process, the vehicle should be unable to move via its own power system. The hydrogen receptacle should be protected with a dust cap, the label beside which should clearly mark the fuel type, nominal operation pressure (NOP) and expiration date of the hydrogen storage tank.

5.3.2.4 Requirements for Hydrogen Pipeline Leakage and Detection

The accessible parts of the fuel pipeline should be tested for hydrogen leakage with more attention paid to the joints. The hydrogen pipeline and test pressure can be classified into two types: The pipelines between the hydrogen storage tank and the fuel cell stack, and the test pressure for this type is set as the actual working pressure for hydrogen leakage detection; the pipelines between the hydrogen receptacle and hydrogen storage tank, and the test pressure for this type is set as 1.25 NOP.

5.3.2.5 Requirements for the Function of Hydrogen Leakage Alarm Device

The hydrogen alarm device shall alert the driver via audible alarm, warning lamp or text display regardless of weather and time. The alarm lamp should be emitted yellow light if the device fails, and the light should be red, when the situation meets any of the first-level alarm conditions as required for in-vehicle hydrogen leakage. It requires to alarm when the hydrogen concentration reaches or exceeds 2.0% ±1.0% during vehicle operation or upon vehicle start. When the hydrogen concentration reaches or exceeds 3.0% ± 1.0%, the alarm state cannot be reset to the normal state until the next restart of the fuel cell engine.

5.3.2.6 Requirements for Hydrogen Discharge of Storage Tank

For the sake of safety consideration with repair and maintenance operation, when the hydrogen system need repair or maintenance after the vehicle putting into use, the FCVs should be equipped with the measures to safely discharge the hydrogen storage tank.

5.4 Hydrogen Leakage and Emission Test

The hydrogen leakage and emission from the FCVs are directly connected with hydrogen safety and energy consumption of the FCVs, and because of special characteristics of hydrogen gas, it is more concerned for the hydrogen safety in confined space such as garage and parking lot located under ground. For the hydrogen emission, its hydrogen concentration caused by emission from tailpipe is directly connected with safety. Dong H. and his colleagues have published a paper [8], which provides deep description for the hydrogen leakage and emission of fuel cell vehicles in confined spaces. And the following parts of hydrogen leakage and emission test are mainly based on the reference [8] with few modifications.

5.4.1 Analysis of Existing Related Standards

The GTR No.13 [9] describes the hydrogen safety of the FCVs in garage and the corresponding safety requirements. And the SAE J2578 [10] also introduces this conditions about hydrogen safety tests of FCVs in the confined spaces. However, about how to test and evaluate this condition, there were no discussion in both GTR No.13 and SAE J2578. In order to provide a solution to this issue, Dr. Dong H. and his research team have developed corresponding test equipment and formed a China-SAE standard [11].

5.4.2 Development of Sealed Test Chamber

The test chamber with suitable internal dimensions, security measures, and air exchange rate was designed and setup to test hydrogen safety of the FCVs in confined space, and which was used to conduct some experimental research.

5.4.2.1 Internal Dimensions

The size of the test chamber is based on the recommendation of the GTR No.13 and SAE J2578. And its internal dimensions should not exceed the length, width, and height of the vehicle by 1000 mm, 1000 mm, and 500 mm, respectively [9]. Basing on the current practical condition of fuel cell passenger cars, the internal size was set as 5650 × 2600 × 1950 mm (L × W × H), which is shown in Figure 5.2.

5.4.2.2 Air Exchange Rate

According to reference materials, the 0.03 ACH (air exchange rate per hour) is selected as the worst-case scenario in general vehicle garages [10, 12]. With all the mechanical ventilation and vents of the test chamber developed are closed, its air exchange rate is less than 0.03 ACH.

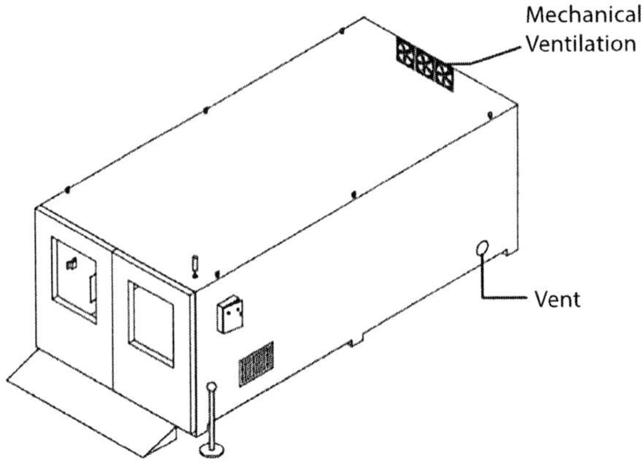

Figure 5.2 Sealed test chamber for FCV hydrogen safety test (reprinted with permission [8]).

And when its two vents are opened and two fans are turned on, its air exchange rate is 6 ACH. With all ventilation fans turned on and vents opened, its air exchange rate is 9 ACH, which is for the emergency condition to exchange internal hydrogen to out space quickly.

5.4.2.3 *Security Measures Adopted for Test Chamber*

Firstly, the developed test chamber was designed and configured with complete explosion-proof parts, which include lamps, fans, cables, and other related components. And it was well grounded for elimination of static electricity. Secondly, the hydrogen sensors were used to monitor H2 concentration of test chamber. When the hydrogen concentration reaches dangerous level set according to the standard, the audible and visual alarm will be initiated and the emergency ventilation is automatically activated to expel the hydrogen inside.

5.4.2.4 *Arrangement of Key Components*

To measure hydrogen concentration in the test chamber, the hydrogen sensors with range of 0-10000 ppm are mounted at each internal corners of the enclosed compartment. And the data collection system can sample the hydrogen concentration values with frequency no less than 1 Hz.

Three automatic controlled electric fans were installed near the ceiling of test chamber, with the air exchange rate of 86.0 m^3/h with 1600 rpm for each. Two other circulation vents were located near the ground floor. With fans and vents are all closed and turned off, the air exchange rate is less than 0.03 ACH. When two fans operated and two vents are opened or turned on, the air exchange rate can reach 6 ACH.

5.4.3 Test Conditions

Two scenarios were set to conduct the experimental investigations:

1) H2 leakage case with air exchange rate of 0.03 ACH under parking condition;
2) H2 emission case with air exchange rate of 6 ACH under short time combined operating conditions.

5.4.4 Test of Two Fuel Cell Passenger Cars

Two fuel cell passenger cars (thereafter, they are referred as vehicle A, vehicle B respectively) were used to conduct H2 leakage and hydrogen emission tests. The nominal working pressure (NWP) of these two test FCVs is 70 MPa, their H2 tank were both mounted under the rear seats, vehicle A adopts Type III hydrogen storage tank, and vehicle B adopts Type IV hydrogen storage tank.

5.4.5 Test Results Analysis

5.4.5.1 Hydrogen Leakage in the Parking State

The test results of hydrogen leakage in parking state of vehicle A and vehicle B were shown in Figures 5.3 and 5.4 respectively, which show different levels of hydrogen leakage. After vehicle A parked for 0.2 h and vehicle B parked for 0.5 h, H2 leakage were detected in the test chamber. With the time going on, the hydrogen concentration increased linearly. After 8 h, the highest hydrogen concentration in the chamber for vehicle A and vehicle B were 125 ppm and 42 ppm respectively with Vehicle B has a lower hydrogen leakage value than that of vehicle A.

Judged from the test results, the test period of 8 hours is enough to examine the hydrogen leakage of the vehicle for daily using condition.

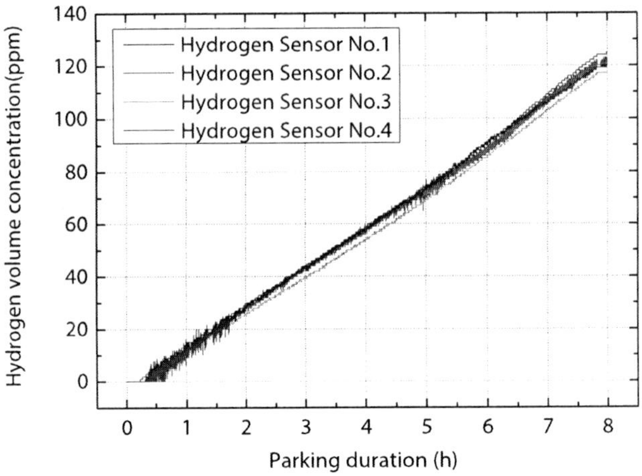

Figure 5.3 Hydrogen concentration changing in the parked state for vehicle A (reprinted with permission [8]).

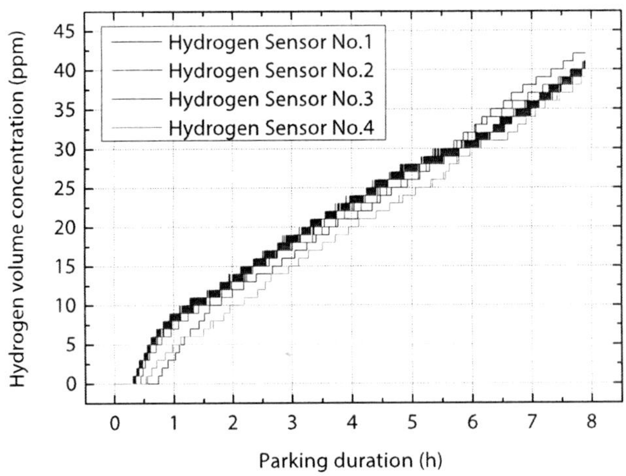

Figure 5.4 Hydrogen concentration changing in the parked state for vehicle B (reprinted with permission [8]).

Taking the sensor of No. 1 position for vehicle B as an example, if its H2 concentration is increased at the measured rate, it would need about 1800 h to reach the defined safety concentration limit of 1%.

For the test, the internal volume of the test chamber confined space minus the volume of the vehicle is V_{Room} (L). At time t, the rate of gas entering chamber G_g (L/min) is the sum of the hydrogen leakage rate Q_{H2}(L/min) and the fresh air rate Q_{air}(L/min). In this case, the H2 concentration in test chamber can be expressed by the Equation (5.1):

$$G_{H2} = \frac{Q_{H2}}{Q_{H2} + Q_{air}} \tag{5.1}$$

After test duration Δt, the volume of gas entering the chamber is calculated by Equation (5.2):

$$G_g = (Q_{H2} + Q_{air})\Delta t \tag{5.2}$$

The hydrogen concentration in the chamber is C_{H2}. The variation of C_{H2} can be expressed by the Equation (5.3):

$$G_g G_{H2} + (V_{Room} - G_g \Delta t)C_{H2} = V_{Room}(C_{H2} + \Delta C_{H2}) \tag{5.3}$$

With combination of Equation (5.1), (5.2) and (5.3), Equation (5.4) was deduced as following:

$$\frac{Q_{H2}}{Q_{H2} + Q_{air}}(Q_{H2} + Q_{air})dt + [V_{Room} - (Q_{H2} + Q_{air})dt]C_{H2} = V_{Room}(C_{H2} + dC_{H2}) \tag{5.4}$$

As a consequence, the H2 concentration in test chamber, C_{H2}, can be expressed as Equation (5.5):

$$C_{H2}(t) = \frac{Q_{H2}}{Q_{H2} + Q_a} - \frac{Q_{H2}}{Q_{H2} + Q_a} e^{-\left(\frac{Q_{H2}+Q_{air}}{V_{Room}}\right)t} \tag{5.5}$$

Where, the air flow in/out of the confined space is Q_{air}(L/min) which is calculated by Equation (5.6):

$$Q_{air} = \frac{R_{air} * V_{Room}}{60} \tag{5.6}$$

Where, $C_{H2}(t)$ is hydrogen concentration in the confined space, Q_{H2} is the total hydrogen discharge rate from the vehicle, R_{air} is the air exchange

Table 5.4 Estimation results of hydrogen leakage of vehicle A and B (reprinted with permission [8]).

Items	Vehicle A	Vehicle B
Total hydrogen leakage rate (mL/min)	6.43	2.14
The rate of hydrogen permeation through container (mL/min)	2.35	1.80
The rate of hydrogen leakage from valves, pipelines and joints (mL/min)	4.08	0.34
The percent of hydrogen from container (%)	36.5	84.1
The percent of hydrogen from valves, pipelines and joints (%)	63.5	15.9

rate of the confined space, V_{Room} is the volume of the confined space minus the material volume of the vehicle.

As mentioned above, the total hydrogen discharged by FCV was mainly due to the hydrogen permeation through hydrogen tank and hydrogen leakage from the high-pressure parts such as valves, pipelines and joints.

Based on the data provided by the manufacturers, the hydrogen permeation rates of the hydrogen storage tank of vehicle A and B were approximately 2.35 mL/min and 1.80 mL/min respectively. With this method, the hydrogen leakage characteristics of vehicles could be quantitatively evaluated, as shown in Table 5.4.

5.4.5.2 *Hydrogen Emissions Under Combined Operating Conditions*

As shown in Figure 5.5, vehicle A entered the start-up purge process after start-up. And the hydrogen concentration from the No. 3 sensor rapidly increased to 695 ppm because of its position near the vehicle tailpipe. Subsequently the hydrogen near No. 3 sensor diffused quickly in the chamber and was detected by sensors of No. 1, No. 2, and No. 4. For this test, the fuel cell engine should be maintained its idle state for 10 minutes.

During idling condition, the H2 concentration at each position increased very slowly. Subsequently, with the shutdown operation, the FCE entered the shutdown purge state. The hydrogen concentration at the No. 3 position rapidly increased from 242 to 2356 ppm. Meanwhile, the hydrogen concentration at other positions also increased. Finally, because of the

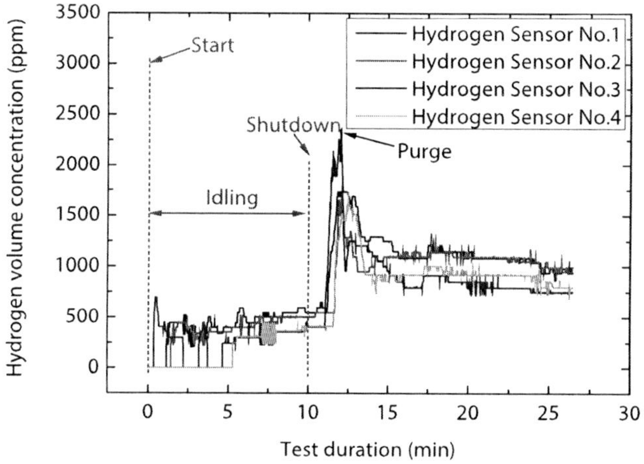

Figure 5.5 In-chamber hydrogen concentration under combined operating conditions for vehicle A (reprinted with permission [8]).

operation of mechanical ventilation, the H2 concentration decreased gradually. During the test for vehicle A, the shutdown purge process caused the highest hydrogen concentration value occurred at the No. 3 sensor which was near the tailpipe of vehicle A.

As shown in Figure 5.6, vehicle B also entered the start-up purge process after start-up, which caused hydrogen discharged from the tailpipe.

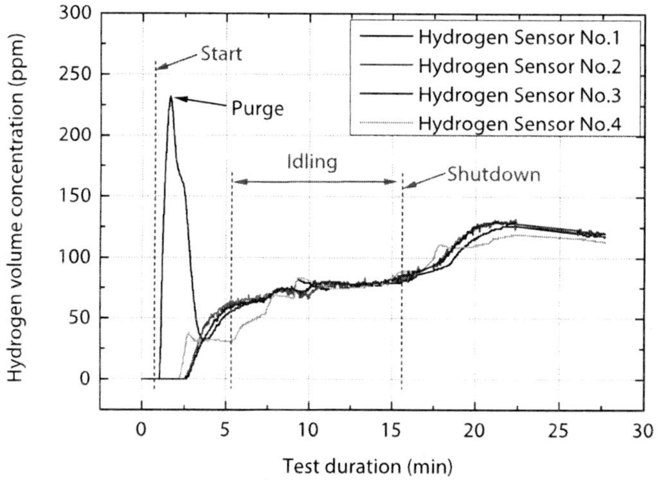

Figure 5.6 In-chamber hydrogen concentration under combined operating conditions for vehicle B (reprinted with permission [8]).

As a consequence, the hydrogen concentration at the No. 1 sensor rapidly increased to 232 ppm because its position was near the outlet of tailpipe. Then the hydrogen near position No. 1 diffused to other positions quickly and was detected at positions No. 2 to No. 4. Then, the FCE was restarted after 6 minutes and entered idle state which lasted until the 16th minute.

During idle process, the hydrogen concentration at each position increased very slowly and remained at approximately 75 ppm. At the 16th minute, the vehicle was shut down and the fuel cell engine entered its shut-down purge process. This process caused hydrogen concentration in the chamber increased to 130 ppm. Subsequently, because of mechanical ventilation activation, the hydrogen concentration decreased gradually.

5.5 Test for Energy Consumption and Range of FCVs

The ISO 23828-2013 and SAE J 2572-2014 both describe the hydrogen consumption test using the temperature and pressure method, gravimetric method, and flow rate method [13, 14], the method defined by these two standards can meet the energy consumption test of the FCVs with bigger fuel cell engine as main power and smaller traction battery as supplement. It is obviously that these three methods cannot meet the test requirements for the electric energy consumption and range test of plugin hybrid fuel cell vehicles (FCVs), and the plugin hybrid configurations are the main technology routs which are used currently by the FCVs in China, especially the fuel cell city bus and heavy duty truck. A new test method for the hydrogen consumption, electric energy consumption, and range of FCVs is proposed without using of energy consumption correction, which can improve the operability of the test and avoid the conversion between electric energy and hydrogen.

Taking into consideration of the different configurations of power trains, the FCVs can be classified into two main types which are the plugin FCVs and the non-plugin FCVs. And the range test is conducted using the chassis dynamo meter after vehicle preparation with necessary data collected during the test; and after completion of the test, the data are calculated according to prescribed steps for obtaining hydrogen consumption and electric energy consumption. The details of this new method are discussed on the paper [15] by Dr. Dong H. and *et al.*, which described the test and calculation process in detail. The following parts of test for energy consumption and range of FCVs are based on reference [15] with few modifications.

5.5.1 Test Vehicle Preparation

Before the test, the test vehicle should be prepared to meet the following requirements:

1) Refuel the vehicle to ensure the SOC of hydrogen storage tank reaching 99%±1% to make it in full hydrogen charged state.
2) For the plugin FCVs, charge the traction battery to 100% SOC; for the non-plugin FCVs, adjust the SOC of traction battery to the full state or other value set by the vehicle manufacturer.
3) Then soak the vehicle at 25±5°C for at least 12 h.
4) After completion of soaking, move the vehicle onto the dynamometer and restrain it according to the test requirements.

5.5.2 Test Procedure

The test procedure should meet the following requirements:

1) The vehicle shall be loaded firstly as requirements according to vehicle classification, and then be started according to the procedure specified by the vehicle manufacturer and the appropriate driving cycle should be adopted until the end of test.
2) The time duration for stop cases between each drive cycle should not be more than 3 times, and the total stop time duration should not be more than 15 minutes.
3) For the plugin FCVs, the traction battery (or other REESS) should be charged within 30 min after the end of the test according to the specifications, and the electric energy from the grid, E in Wh, should be recorded, and E is the electric energy consumed by the test vehicle.
4) After the vehicle soaked with 25±5 °C for at least 8 h, the vehicle is moved (without using vehicle power) to the hydrogen refuel station to refill its hydrogen storage tank to the full state (99%±1%), and the mass of hydrogen refilled into vehicle should be recorded as $m_{HYDROGEN}$ in kg.
5) All related data required for evaluation should be collected during the test.

5.5.3 Requirements for Data Collection

During the test, the following data should be collected:

1) The distance covered by the vehicle from the start to the end of the test is recorded as D in km, and the time duration is recorded as T in s.
2) The related collected data from the test. For the plugin FCVs, the data are the voltage of traction battery (U_{BAT}), current of traction battery (I_{BAT}), output voltage of fuel cell stack (U_{FC}), output current of fuel cell stack (I_{FC}); For the non-plugin FCVs, the U_{BAT} and I_{BAT} should be collected.
3) After the test, the refueled hydrogen mass ($m_{HYDROGEN}$ in kg) is the total mass of the hydrogen consumed during the test. And for the plugin FCVs, the recharged electric energy (E in kWh) charged from the grid is the total electric energy consumed during the test.

5.5.4 Range and Energy Consumption Calculation for FCVs

5.5.4.1 *Data Process Steps for the Plugin FCVs*

For the plugin FCVs, the data should be analyzed according to the following steps:

1) Basing on the collected data of U_{FC}, I_{FC} and T, the total output energy, E_{FC}(kWh), from the fuel cell stack is calculated by Equation (5.7):

$$E_{FC} = \frac{\int_0^T I_{FC} U_{FC}\, dt}{3600 \times 1000} \tag{5.7}$$

2) Based on the collected data of U_{BAT}, I_{BAT} and T, the net energy change of traction battery(REESS), E_{BAT}(kWh), is calculated by Equation (5.8):

$$E_{BAT} = \frac{\int_0^T I_{BAT} U_{BAT}\, dt}{3600 \times 1000} \tag{5.8}$$

3) The total energy, E_D (kWh), from the traction battery (REESS) and the fuel cell stack is calculated by Equation (5.9) with the addition of E_{FC} and E_{BAT}:

$$E_D = E_{FC} + E_{BAT} \tag{5.9}$$

4) Based on the results obtained above, the percentage (η_{FC}) of E_{FC} over E_D and the percentage (η_{BAT}) of E_{BAT} over E_D are calculated by Equations (5.10) and (5.11) respectively:

$$\eta_{FC} = \frac{E_{FC}}{E_D} \times 100\% \tag{5.10}$$

$$\eta_{BAT} = \frac{E_{BAT}}{E_D} \times 100\% \tag{5.11}$$

5) Based on D and η_{FC}, the range contributed by fuel cell stack, D_{FC} (km) is calculated by Equation (5.12) :

$$D_{FC} = D \cdot \eta_{FC} \tag{5.12}$$

6) Based on D and η_{BAT}, the range contributed by the traction battery, D_{BAT} (km), is calculated by Equation (5.13):

$$D_{BAT} = D \cdot \eta_{BAT} \tag{5.13}$$

7) Based on the measured data $m_{HYDROGEN}$ and D_{FC}, the hydrogen computation rate, $C_{HYDROGEN}$(kg/100km), is calculated by Equation (5.14):

$$C_{H_2} = 100 \times \frac{m_{H2}}{D_{FC}} \tag{5.14}$$

8) Based on the measured data E and the calculated D_{BAT}, the electric energy computation per 100 km, C_E(kWh/100km), is calculated by Equation (5.15):

$$C_E = 100 \times \frac{E}{D_{BAT}} \tag{5.15}$$

Based on the above calculation steps, the fuel economy of the plugin FCVs can be described with the total range, ranges contributed by traction battery and hydrogen respectively, hydrogen consumption rate, and electric energy consumption rate without energy conversion between electricity and hydrogen.

5.5.4.2 *Data Analysis for the Plugin FCVs*

For the non-plugin FCVs, the percentage of the net energy change of traction battery (REESS) over the total hydrogen energy should be calculated firstly. If this percentage is less than 3%, the contribution of the traction battery to the range can be neglected. Otherwise, the SOC of traction battery should be adjusted to conduct another test to meet this percentage limit requirement of less than 3%. The whole calculation process is as following:

1) With collected data of U_{BAT}, I_{BAT} and T, the net energy change of REESS, E_{BAT} (kJ), is calculated by Equation (5.16):

$$E_{BAT} = \left| \int_0^T I_{BAT} U_{BAT} \, dt / 1000 \right| \tag{5.16}$$

Compared with that of plugin FCVs, the net energy change of traction battery (REESS) in the non-plugin FCVs is smaller, and this value can be negative or positive. Therefor the absolute value is used in the following calculation.

2) With total hydrogen mass, $m_{HYDROGEN}$, and the lower heating value of hydrogen, the total energy from hydrogen consumed in the range test, $E_{HYDROGEN}$ (kJ), is calculated by Equation (5.17):

$$E_{H_2} = m_{H2} \cdot \mathrm{LHV_{H2}} \tag{5.17}$$

Where, the $\mathrm{LHV_{HYDROGEN}}$ is the lower heating value of hydrogen, which is 1.2×10^5 kJ/kg.

3) The calculated results from Equation (5.16) and (5.17) are then used to calculate the percentage, $\eta_{EBAT/EHYDROGEN}$, of net energy change of the traction battery (REESS) over the total energy of hydrogen:

$$\eta_{EBAT/EH2} = \frac{E_{BAT}}{E_{H2}} \times 100\% \tag{5.18}$$

4) If the $\eta_{EBAT/EHYDROGEN} \leq 3\%$, the contribution of traction battery is neglected, then directly calculate the hydrogen consumption per 100 km, $C_{HYDROGEN}$ (kg/100km), using Equation (5.19):

$$C_{H2} = 100 \times \frac{m_{H2}}{D} \tag{5.19}$$

Based on these above steps, the economy performance of the non-plugin FCVs can be presented using the range which is measured directly and hydrogen consumption per 100 km from calculation.

5.5.5 Test of Range and Energy Consumption for Fuel Cell Passenger Car

To verify the new method, one plugin fuel cell passenger car and one non-plugin one were tested and analyzed basing on the above mentioned process.

5.5.5.1 *Test of Plugin Fuel Cell Car*

The test vehicle was a passenger car with a curb weight of 2010 kg, and the vehicle was charged to full state for traction battery and refueled to 100% SOC of the hydrogen storage tank. The range and energy consumption test were conducted basing on the drive cycle of New Europe Drive Cycle (NEDC), and the voltage and current of both fuel cell stack and traction battery were measured over the whole test process. When the range reached 385.01 km, the indicator in the instrument panel informed the driver to stop the vehicle, and the test ended.

According to the data calculation methods described above, and basing on the five values of range, output energy of fuel cell stack, net energy change of traction battery, mass of hydrogen consumption, and the electric energy from grid, another five values were calculated which are the total energy from the traction battery and fuel cell stack, contribution

Table 5.5 Results of hydrogen consumption and range of a plugin fuel cell car (reprinted with permission) [15].

Items	Data	Remark
range (km)	385.01	measured
output energy of fuel cell stack (kWh)	65.6	measured
net energy change of traction battery (kWh)	16.7	measured
mass of HYDROGEN consumption (kg)	3.55	measured
electric energy from grid (kWh)	17.3	measured
total energy (kWh)	82.3	calculated
energy contribution of traction battery (%)	20.3	calculated
energy contribution of fuel cell stack (%)	79.7	calculated
range contributed by traction battery (km)	78.1	calculated
range contributed by fuel cell stack (km)	306.9	calculated
hydrogen consumption (kg/100 km)	1.16	calculated
electric energy consumption (kWh/100 km)	22.1	calculated

percentage of traction battery, contribution percentage of fuel cell stack, range contributed by traction battery, and range contributed by fuel cell stack. The related values are listed in Table 5.5.

5.5.5.2 *Test of Non-Plugin Fuel Cell Car*

A non-plugin fuel cell car with curb weight of 1950 kg was also tested using this new method. The hydrogen storage tank of the test car was refueled to its full state and the SOC of its traction battery was adjusted to the value set by manufacture. The driving cycle of NEDC was adopted to conduct the range test with measurement of voltage and current of traction battery. When the indicator informed the driver to stop the vehicle, the measured range was 450.03 km.

The test results are listed as following: The total hydrogen consumed was 4.55 kg with the total energy of hydrogen consumed in the test was 546, 000 kJ. The net energy change of traction battery was 4, 320 kJ (1.2 kWh), and the percentage was 0.79% for the net energy change of traction battery over the total consumed hydrogen energy. This percentage value of 0.79%

means that the range contributed by traction battery was only 3.65 km (calculated by 0.79 %×450 km) which was neglected because that the condition of $\eta_{EBAT/EHYDROGEN} \leq 3\%$ was met. The hydrogen consumption is 1.01 kg/100km which was directly calculated with electric energy consumption neglected.

5.5.6 Test of Range and Energy Consumption for Fuel Cell Truck

A fuel cell truck (FCT) was tested for its range, energy consumption, and hydrogen emission under two other different driving cycles, which are the constant speed of 40 km/h and the China heavy-duty commercial vehicle test cycle for heavy truck (CHTC-HT) [16]. The following discussion of test of range and energy consumption for fuel cell truck is mainly based on reference [17] with some modifications.

5.5.6.1 Brief Introduction of Test Vehicle and Test Cycles

In this study, the power train of the fuel cell truck was equipped with two electric power sources, which are one fuel cell engine and one traction battery with rated power of 46 kW and 53 kW respectively. According to the control strategy, the FCT is only driven by the power from the traction battery in the traction battery mode with its SOC within setting range; and when in the hybrid mode, the FCT is powered by combination of the fuel cell engine and the traction battery basing on the practical power requirements, vehicle speed, and the state of charge (SOC) of traction battery. The hydrogen system in the test FCT is consisted of three hydrogen storage tanks with total volume of 420 L and NOP of 35 MPa.

The CHTC-HT drive cycle is defined by China national standard for heavy-duty trucks with a total vehicle mass greater than 5, 500 kg. The duration of one complete CHTC-HT cycle is 1800 s, and can be divided into three different parts which represent three different road conditions of urban, suburban, and highway, with time percentage of 19.0%, 54.9%, and 26.1% of the CHTC-HT duration, respectively. The detail data of CHTC-HT drive cycle can be found in reference [16].

5.5.6.2 Test Requirements

The test of range, energy and emission was completed on the heavy-duty chassis dynamometer with constant speed of 40 km/h and CHTC-HT drive cycle respectively. The traction battery and hydrogen storage tanks of

FCT were recharged and refueled to their top SOC states before test. And there are several test requirements for test cycle which are speed tolerance, time deviation limit, and data collections.

For the constant speed of 40 km/h, the FCT on the dynamometer should maintain the constant speed of (40 ± 2) km/h. During the test, the SOC of hydrogen storage tank was reduced gradually to its lower limit of 8% (set by manufacture), at this point, the FCT may be switched to the battery mode depending on the SOC of its traction battery. And the whole test would come to the end when the vehicle speed was reduced and less than 36 km/h with its hydrogen and electric energy cannot supply enough energy for continuous driving with required vehicle speed.

For the CHTC-HT cycle, the speed deviation during the test should not exceed ± 3 km/h with time deviation no longer than 2 s each time, and the total of deviation time should not longer than 10s during the test. When the SOC of hydrogen storage tank was reduced to 8%(set by manufacture), the FCT may be immediately switched to the traction battery mode, and the test continued until the deviation of vehicle speed was beyond the range of ±3 km/h or the total time deviation exceeded 10s.

During the test, the relevant data were collected which include the current, voltage of the fuel cell stack and the traction battery, range distance and hydrogen mass flow rate.

5.5.6.3 Power Change and Energy Consumption Results

With the constant speed of 40 km/h, the Fuel cell stack output power changed between 13.97 kW and 14.19 kW, which was nearly constant power output for the fuel cell stack. And its traction battery operated in the lower discharge condition with output power changed between 0.36 and 1.06 kW. With the CHTC-HT cycle, the Fuel cell stack output power was from 11.69 to 40.94 kW with bigger variation. The output power of the traction battery is between 0 and 43.59 kW with maximum charging power up to 128.02 kW during regenerative braking process. During one cycle of CHTC-HT, the output power of traction battery and fuel cell stack at two driving cycles were shown in Figures 5.7 and 5.8 respectively.

Basing on the collected data and calculation, the final results for the range, electric energy consumption, and hydrogen consumption are as followings: with the constant speed of 40 km/h, the range, electric energy consumption, and hydrogen consumption are 561 km, 36.85 kWh/100km, and 1.98 kg/100km respectively; with the CHTC-HT drive cycle, the corresponding values are 317 km, 57.09 kWh/100km, and 3.01 kg/100km respectively.

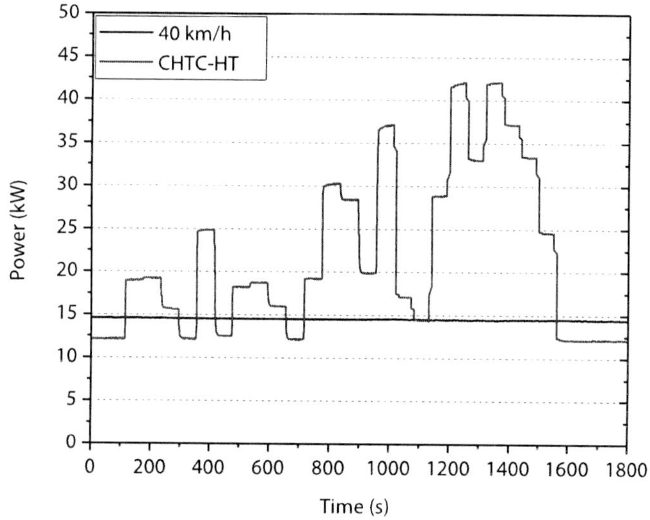

Figure 5.7 Fuel cell stack power change under two different driving cycles (reprinted with permission [17]).

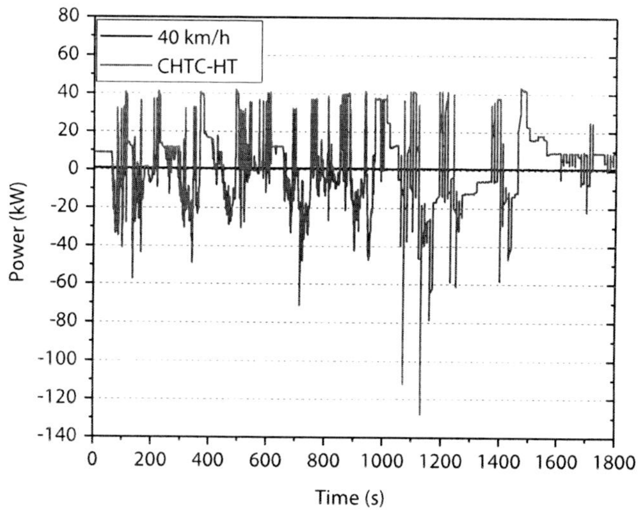

Figure 5.8 Battery power change under two different drive cycles (reprinted with permission [17]).

5.5.6.4 *Hydrogen Emission and Hydrogen Leakage*

The test of hydrogen emission and leakage for the vehicle parked in the confined space has been discussed above. But for total emission and leakage in drive cycle condition, its test and calculation methods are different. During the range test, the hydrogen mass flow rate was measured, and the actual hydrogen consumed by the fuel cell truck is calculated by Equation (5.20):

$$M_{H2} = \int_0^T m_{H2}\,dt \tag{5.20}$$

Where, M_{H2} is actual hydrogen consumption (g); m_{H2} is the measured hydrogen flow rate (g/s); T is test duration (s).

Actually, not all the hydrogen supplied form hydrogen storage tank was participated the reaction with oxygen of air. The theoretical hydrogen consumption of the fuel cell stack can be determined by Equation (5.21):

$$M'_{H2} = \int_0^T m'_{H2}\,dt \tag{5.21}$$

Where, M'_{H2} is theoretical hydrogen consumption of the fuel cell stack (g); m'_{H2} is theoretical hydrogen mass flow rate of the fuel cell stack (g/s), the theoretical hydrogen mass flow rate is determined by Equation (5.22):

$$m'_{H2} = \frac{mIN}{2F} \tag{5.22}$$

Where, m is the Mole mass of hydrogen, 2.016 g/mol; I is the current of the fuel cell stack (A); N is the number of fuel cell units in the stack; F is the Faraday constant, 96485 C/mol.

Thus, the mass of hydrogen emission (including leakage) in the test duration T can be calculated by Equation (5.23):

$$M''_{H2} = M_{H2} - M'_{H2} \tag{5.23}$$

Where, M''_{H2} is the total hydrogen emission mass during the test duration T.

The hydrogen emission per 100 km is calculated by Equation (5.23):

$$C_{H2} = \frac{100 \times M''_{H2}}{D} \tag{5.23}$$

Where, C_{H2} is the hydrogen emission per 100 km (g/100km).

Hydrogen emission per kWh can be determined by Equation (5.24):

$$b_{H2} = \frac{3600 \times M''_{H2}}{Q_s} \tag{5.24}$$

Where, b_{H2} is the hydrogen emission per kWh (g/kWh); Q_s is the total output energy (kJ) of the fuel cell stack during the test duration T, which is calculated by Equation (5.25):

$$Q_S = \int_0^T P_S \, dt \tag{5.25}$$

Where, P_s is the output electric power of the fuel cell stack.

Basing on the measured data and calculations according to the equation (5.20) to equation (5.25), the corresponding results are as follows: For the test of constant speed of 40 km/h, the hydrogen emission per 100 km (g/100km) and the hydrogen emission per kWh (g/kWh) are 89.13 and 241.69 respectively; for the test of CHTC-HT drive cycle, the corresponding values are 2.93 and 3.72 respectively. It is obviously that the hydrogen emission per 100 km and hydrogen emission per kWh with the CHTC-HT drive cycle is higher than those of the constant speed of 40 km/h, which is mainly because of its varying power requirements and frequent speed variation during the test.

5.6 Subzero Cold Start Test for FCVs

Cold start performance under subzero temperature is one of the key performance of the FCVs. The water produced in fuel cell stack can get frozen under subzero conditions, which cause the start of FCE become difficult and require the special control strategy to get start successfully. Because of the difference between the fuel cell vehicles and internal combustion engine vehicles, a special method is needed to test and evaluate the subzero start for fuel cell vehicles. A standard of China-SAE was set up to meet the standard need of subzero cold start test and evaluation for the FCVs, which is the T/CSAE122-2019< Test methods for cold start performances of fuel cell electric vehicles under subzero temperature> [18]. The following parts of test method for cold start under subzero temperature are mainly based on the standard of T/CSAE122-2019, the reference [18].

5.6.1 Test Method for Cold Start Under Subzero Temperature

The main parts of the T/CSAE 122-2019 include test conditions, test method, and test results, which were described as followings.

5.6.1.1 Test Conditions

Firstly, the test vehicle should keep the conditions or meet the requirements of manufacturer such as structure sizes and technical parameters, viscosity of lubrication oil, coolant, and hydrogen fuel.

Secondly, the subzero temperature soaking should be controlled at the target value with tolerance of ±2°C, and should not be greater than 0°C.

Thirdly, for safety consideration, the test chamber should be equipped with hydrogen alarm sensors and forced ventilation system.

5.6.1.2 Vehicle Soaking Under Subzero Temperature

The FCVs should be soaked for enough time according to the following steps:

1) Before soaking, it is allowed to adjust vehicle conditions and SOC of traction battery or other REESS according to manufacture instructions and record the SOC and time used for adjusting.
2) The temperature of test chamber should be reduced to the required target value.
3) If the measures were taken by the FCV for heating fuel cell engine, this function should stop working before soaking.
4) During the temperature decreasing process, it is allowed to conduct the start and stop operation only once with total duration not greater than 5 minutes (the purge time is excluded).
5) The FCVs should be kept under the target temperature for at least 12 hours from the beginning when the temperature in test chamber reaches the target value.

5.6.1.3 *Test Process for Subzero Cold Start of FCE*

From the start operation to the end of test, the data should be collected which include the voltage and current of fuel cell stack, the voltage and current of traction battery or other REESS, the hydrogen concentration of the emission from tailpipe. And during the test, the requirements for the hydrogen concentration of emission should meet the standard such as GTR No.13 and SAE J2578, which are that the 3-second rolling average should not exceed 4% and the instantaneous concentration should not exceed 8%.

The following steps should be followed to conduct this test:

1) After soaking completion, start the FCE according to the instruction of vehicle manufacture.
2) After start operation beginning, the air conditioner can be turned on to consume electric power from FCE.
3) Record the duration from beginning to the time when the vehicle show its ready state (the label of "READY" or "OK" is lighted in instrument panel), this time duration is represented as t1;
4) Record the time from beginning to the point when the output power of fuel cell stack reaching 1.0 kW, this time duration is represented as t2, and the t2 is referenced as the FCE start time. After the power output of FCE reaching the 1.0 kW, the FCE should continually operate for at least 10 minutes with the output power no less than 1.0 kW.
5) After 10 minutes, the FCE should be shut down following the instruction of manufacture.
6) If these above steps are completed smoothly, it means that the FCE can start successfully under the setting subzero temperature; if some faults occurred or the FCE stopped by itself during this process, it means that the start process fails and need retest from soaking process as the first step.

During the test, the hydrogen concentration of the emission from tailpipe should also meet the standard requirements mentioned above.

5.6.1.4 *Test Process for Subzero Cold Start of FCVs*

In some conditions, the FCE starting successfully does not mean that the FCV can launch or move successfully, therefore another test process is needed to determine whether the FCV can move as required. The test is conducted using dynamometer in the laboratory with load and road resistance set to simulate practical conditions, and the road resistance setting should be able to reflect the subzero temperature influence.

1) The test vehicle should be soaked firstly according to the requirements;
2) After completion of soaking, start the vehicle following the steps suggested by vehicle manufacture;
3) When the label of "READY" or "OK" in the instrument panel is lighted, record the time duration as t3. This state means that the vehicle drive system is ready to move the vehicle;
4) When the vehicle reaches ready state, the acceleration pedal should be depressed completely until the output power of fuel cell stack reaches 50% of the rate power of fuel cell engine, record the time as t4;
5) Then release the acceleration pedal, and depress the brake pedal to stop the vehicle in one minute.
6) Within 3 minutes after stop, the vehicle should continuously operate on dynamometer to finish one complete drive cycle such as the NEDC or other drive cycles. And vehicle speed and time deviation should also meet standard requirements during the corresponding drive cycle.
7) Then turn off the vehicle to end the test according to steps suggested by manufacture;
8) If all these above steps were finished smoothly, it means that the vehicle can start at subzero temperature successfully. Otherwise, it suggests that the start process fails. And should be retested from beginning.

5.6.1.5 Data Collection and Results

These following data should be measured and recorded from beginning: the voltage and current of fuel cell stack, the voltage and current of traction battery, the hydrogen concentration of emission. And the sample time should cover the whole test process with sample frequency less than 5Hz for electric and hydrogen concentration signals.

With the data collected, the energy consumed from traction battery (or other REESS) and hydrogen can be calculated. And the start time duration can be determined. Then energy consumption and time are used to evaluate start performance of FCVs.

5.6.2 Test for Subzero Cold Start of FCVs

For the test subzero cold start for several FCVs, a special test system was developed and used to test several FCVs for verification, and the following descriptions are based on reference [19] with some modifications.

5.6.2.1 Test System Development

Because of the difference of subzero cold start conditions between the FCVs and the conventional vehicles, the vehicle soaking time, judgement of start status, vehicle conditions, test parameters, test devices and test steps were

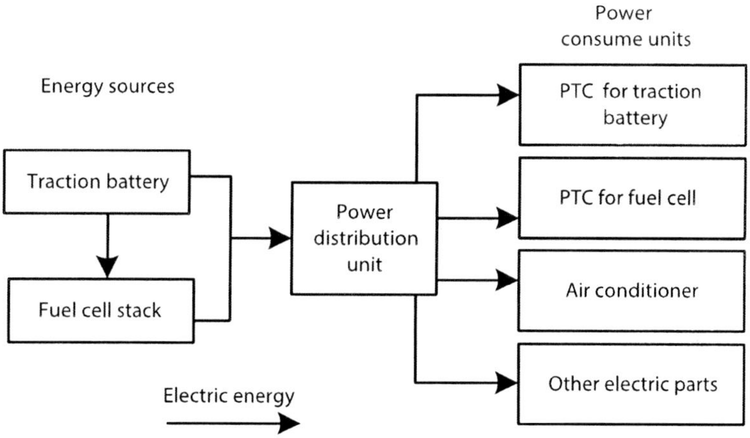

Figure 5.9 Energy flowing schematic of the HCVs in starting process (reprinted with permission [19]).

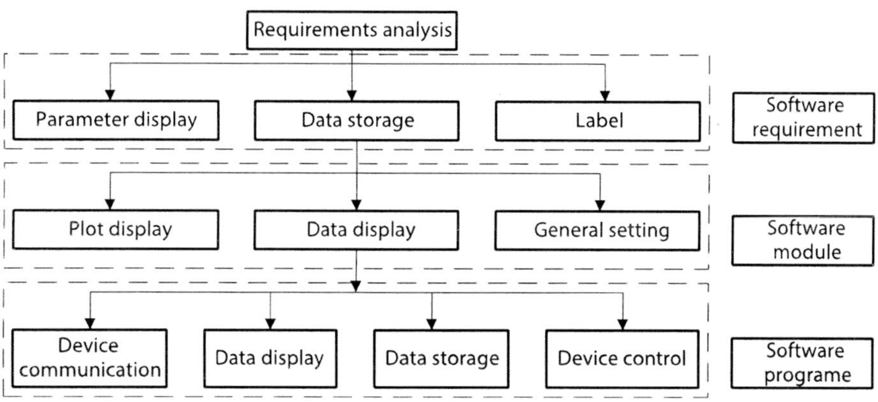

Figure 5.10 Software frame for subzero cold start test of FCVs (reprinted with permission [19]).

defined according to technology features of different FCVs. And a set of special data collection system was developed to meet the requirements of FCVs for the test and evaluation of cold start process with the energy flow schematic in the start process is shown as Figure 5.9 [19]. And the corresponding test device system were developed to measure, display and store the related data, its software frame is shown in Figure 5.10 [19].

5.6.2.2 *Analysis of Test Results*

Several FCVs were tested to verify the new developed test system and subzero start methods. A fuel cell bus was taken as an example to explain the whole test process of cold start performance under - 25°C. The rated power of fuel cell engine for this bus is 30kW. Before subzero cold start test, the fuel cell bus was soaked in the surrounding temperature of -25°C for 24h, then begin to conduct the cold start test. The whole test process was divided into three stages:

1) Early start stage. The battery is discharged to supply its PTC (positive temperature coefficient) heater to warm up itself with constant power, and at the same time the fuel cell stack use its own PTC to warm up itself.
2) Medium stage. The air conditioner system was started to make the fuel cell engine operate at setting output power.
3) Last stage. The start process is completed, the power of fuel cell stack reach 17 kW, and the traction battery changed its working state from discharging to charging.

From the beginning of start to 17kW of out power of fuel cell stack, the total time is 47 minutes. The total energy consumed during start process is the addition of output energy from traction battery and fuel cell stack. The total energy consumed is 11.16 kWh, which was supplied by traction battery and fuel cell stack, they are 9.67kWh and 1.49kWh respectively. The PTCs consumed energy for heating traction battery and fuel cell stack are 5.84kW and 2.31kWh respectively, the air condition system consumed energy is 2.03kw. The power supply and consumption conditions were presented in Figure 5.11 and Figure 5.12 respectively.

During the whole start process, the hydrogen concentration of emission was measured, which was presented in Figure 5.13. From beginning to start of fuel cell stack, the hydrogen concentration cannot be measured without real reaction of hydrogen and oxygen, and that the hydrogen concentration value show that the chemical reaction in the stack is ongoing with maximum value of 6.186%, which is below the standard required value of 8%.

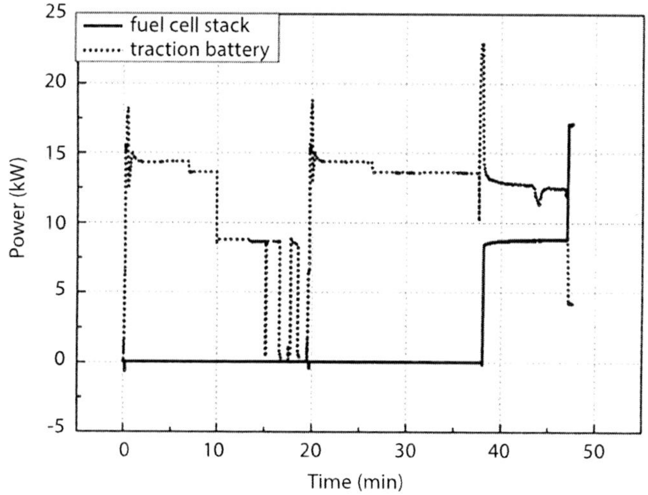

Figure 5.11 Power supplied by fuel cell stack and traction battery (reprinted with permission [19]).

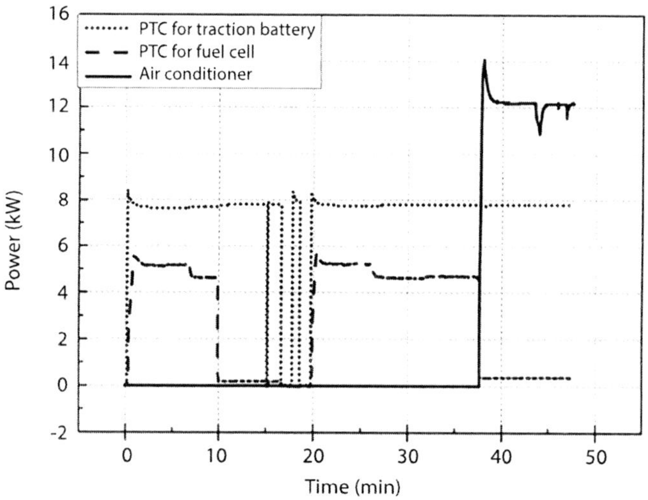

Figure 5.12 Power consumed by air conditioner and PTCs for fuel cell and traction battery (reprinted with permission [19]).

In the test, there were several problems may be occurred:

1) There are hydrogen leakage at subzero temperature, which can cause hydrogen concentration higher than the lower safety limit (10000 ppm) in the test chamber to interrupt the test process;

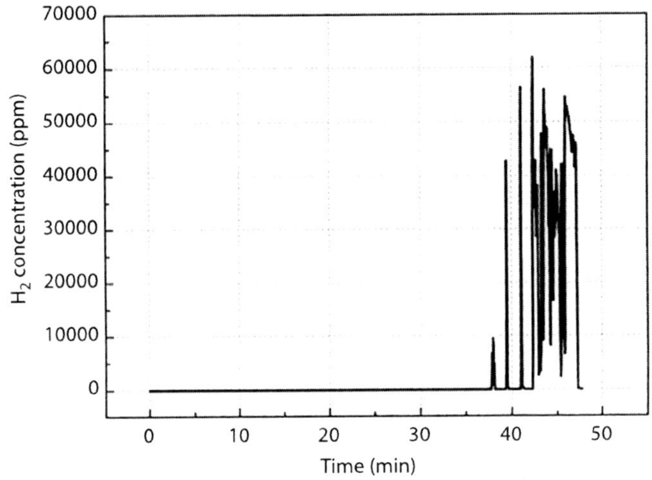

Figure 5.13 Hydrogen concentration of emission from tailpipe during start process (reprinted with permission [19]).

2) There are several cold start fail modes. Some are caused by lower warming or heating capability, some were caused by malfunction of accessory parts at subzero temperature.
3) Energy consumption during cold start process. In some very cold area, some heating or warm keeping systems of FCVs consume electricity energy outside the vehicle such as grid. Therefore how to evaluate the heating energy consumption is still a topic needing further research and discussion.

5.7 Conclusion

In this chapter, the special performance requirements and test methods were briefly discussed, which are about the safety performance test, hydrogen leakage and emission test in confined space, energy consumption and range test, and subzero cold staring test. The discussion in this chapter is trying to provide practical test methods for these above mentioned issues. All the discussions in this chapter were based on the research which were conducted by Dr. Dong H. and his colleagues, and should be recognized as a kind of trial or suggestion for test and evaluation of the fuel cell vehicles.

References

1. Dong, H., Shan, R., Yanyi, Z. *et al.*, Research and application of test and evaluation system for fuel cell vehicles. *IOP Conf. Ser.: Earth Environ. Sci.*, 558, 4, 052058, 2020.
2. SAE J2579-201806. Standard for fuel systems in fuel cell and other hydrogen vehicles. https://saemobilus.sae.org/content/J2579_201806/, April 30, 2020.
3. Fuel cell road vehicle-safety specifications-protection against hydrogen hazards for vehicles fueled with compressed hydrogen, ISO 23273, 2013. https://www.iso.org/standard/64047.html, March 8, 2022.
4. China National Standard GB/T24594-2020. Fuel cell electric vehicles-safety requirements.
5. China National Standard GB 18384-2020. Electric vehicles safety requirements.
6. China National Standard GB 38032-2020. Electric buses safety requirements.
7. China National Standard GB 38031-2020. Electric vehicles traction battery safety requirements.
8. Dong, H., Xiaobing, W., Yanyi, Z. *et al.*, Experimental study on hydrogen leakage and emission of fuel cell vehicles in confined spaces. *Automot. Innovation*, 3, 2, 111, 2020.
9. UN ECE.GTR No.13. Global technical regulation concerning the hydrogen and fuel cell vehicles. http://www.unece.org/trans/main/wp29/wp29wgs/wp29gen/wp29glob _ registry. html, March 20, 2018.
10. SAE J 2578-201408. Recommended practice for general fuel cell vehicle safety. https://saemobilus.sae.org / content/J2578_201408/, April 30, 2020.
11. China SAE Standard T/CSAE 123-2019. Fuel cell electric vehicles - test methods and safety requirements for hydrogen leakage and emission in confined space.
12. Adams, P., Bengaouer, A., Cariteau, B. *et al.*, Allowable hydrogen permeation rate from road vehicles. *Int. J. Hydrogen Energy*, 36, 3, 2742–2749, 2011.
13. ISO 23828:2013. Fuel cell road vehicles — Energy consumption measurement — Vehicles fuelled with compressed hydrogen. https://www.iso.org/standard/61031.html
14. J2572_201410. Recommended practice for measuring fuel consumption and range of fuel cell and hybrid fuel cell vehicles fueled by compressed gaseous hydrogen. https://saemobilus.sae.org/content/J2572_201410/, Oct. 10, 2021.
15. Dong, H., Yanyi, Z., Renguang, W. *et al.*, An improved test method for energy consumption and range of fuel cell vehicles. *Int. J. Chem. Eng.*, 1–9, 2020.
16. China National Standard GB/T 28146.2-2019. China automotive test cycle-part 2: Heavy-duty commercial vehicles.
17. Tian, S., Guang, C., Hao, L. *et al.*, Experimental study on fuel economy of fuel cell truck under different driving cycle. *2021 4th Asia Conference on Energy and Electrical Engineering (ACEEE)*, pp. 38–43, 2021.

18. China SAE Standard T/CSAE 122-2019. Test methods for cold start performances of fuel cell electric vehicles under subzero temperature.
19. Shengtao, S., Yongping, H., Dong, H. *et al.*, Development of data acquisition system for freezing start test of fuel cell vehicle. *Battery Bimonthly*, 49, 05, 396, 2019.

6

Hydrogen Production and Polymer Electrode Membrane (PEM) Fuel Cells for Electrical Vehicles

Cigdem Tuc Altaf[1], Tuluhan Olcayto Çolak[2], Alihan Kumtepe[2], Emine Karagöz[2], Ozlem Coskun[2], Nurdan Demirci Sankir[1,2*] and Mehmet Sankir[1,2†]

[1]Department of Materials Science and Nanotechnology Engineering, TOBB University of Economics and Technology, Sogutozu Caddesi Sogutozu, Ankara, Turkey
[2]Micro and Nanotechnology Graduate Program, TOBB University of Economics and Technology, Sogutozu Caddesi Sogutozu, Ankara, Turkey

Abstract

Fuel cells (FC) are regarded as efficient power suppliers for energy storage. They convert chemical energy stored in hydrogen bonds into electrical energy, releasing water as the only by-product. Therefore, it has a great potential for utilizing green energy as diminishing pollutant emission primarily due to the CO_2 released from fossil fuel and the socio-economic dependence on fossil fuel. In the last few decades, fundamental research and advancements in FC technology have improved tremendously, especially accompanying the integration in electrical vehicles (EVs). The most frequently used FCs for vehicle applications is the proton exchange membrane -commonly called polymer electrolyte membrane-FCs (PEMFCs) due to the benefits of low weight, high-power density, and cold-start capability. Using PEMFCs and hydrogen, one can reach about 65% of practical efficiency with no pollutant emission. Today, almost all the major automobile producers are in contact with PEMFC EVs technology. Thus, this chapter emphasizes the analysis of PEMFCs for the applications of EVs. It is designed to combine the assistive research updates on the materials and components to be valuable for the performance enhancement of the FCEV applications. This chapter also brings together the essential data of commercialized PEMFC EVs up to date.

Corresponding author: msankir@etu.edu.tr

Keywords: Electric vehicles, hydrogen, PEM, fuel cell, green energy

6.1 Introduction

6.1.1 Energy Challenges and Green Energy Demand

The expanse of energy consumption all over the world has been gradually increasing in the past century due to the hasty industrialization and growth in population. Thus far, carbon-based fuels have triggered significant damages for the environment due to the CO_2 emission, which is the main cause for climate change and extreme weather incidents and various health problems for living beings. Although the use of carbon-based fuels (coal, natural gas and petroleum) are still prevailing to meet the overall energy demands of the world population, renewable energy sources are being promoted to reduce traditional fossil fuels' consumption for the sake of more sustainable and inhabitable future [1–3]. For instance, the US government grants for energy sector by source reported by International Renewable Energy Agency (IRENA) in 2020 for the years of 2017, 2030 and 2050 (given in Figure 6.1) indicates the gradual decrease in fossil fuel consumption replacing with renewable energy sources.

In this regard, the European Union (EU) has approved establishing the target of renewable energy usage as 10% in transport in 2020 [4]. This target includes hydrogen fuel, green electricity, biofuels, biomethane, etc. Especially, hydrogen is coming into prominence as being free of carbon

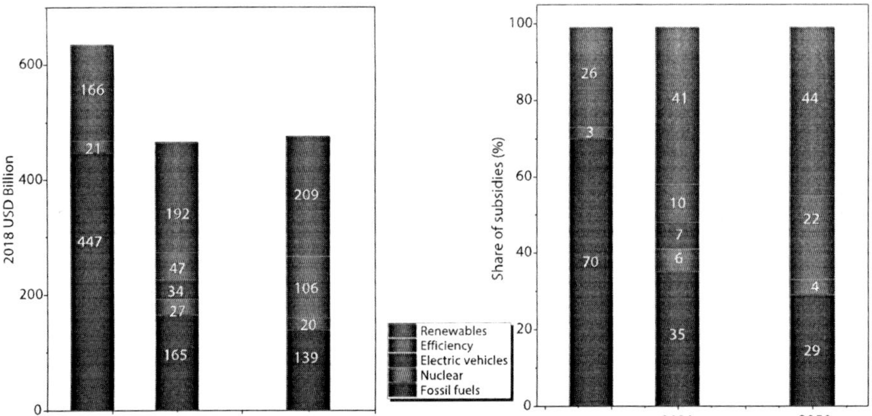

Figure 6.1 The grants for energy sector by US government (Reproduced from the data source: the International Renewable Energy Agency (IRENA) annual book in 2020) [3].

footprint since it releases only water as a by-product after combustion [5]. Hydrogen is an energy carrier (not an energy source), and it is desirable to store renewable energies or energies in chemical bonding for use when the energy is required. Another intriguing research interest is the conversion of hydrogen produced from renewable energy or chemical hydrides into electricity through fuel cell (FC) technology. For instance, hydrogen-powered FCs are found to be highly accomplished to generate power sufficiently for commercialization [6]. In the FCs, the hydrogen and oxygen gases chemically converted into electrical energy and heat along with water as a by-product with no pollutant emission.

Introducing renewable energy sources into the transportation segment is an excellent solution for reducing fossil fuel consumption. In this respect, electric vehicles (EVs) are a new and rapidly developing market which demand electric energy from renewable sources. EVs offer potential advantages of significant efficiency expansions and reduce fossil fuel dependence [4]. For instance, The Republic of Korea's goal has been announced as the launch of 850k hydrogen-fuel and 3 million electrical vehicles until the year 2030 [4]. Since the first hydrogen fuel cell vehicle (FCV) concept debuted in Osaka in the late 90s, car manufacturers have turned towards FC technology. As Japanese foremost car manufacturers, Toyota and Honda are leading in the effective hydrogen FCEV market [7]. By the government of Japan, the greenhouse gas emissions are expected to drop by 14% with the electricity mix transition until 2030 [7, 8].

6.1.2 FC in Green Energy Aspect

As mentioned before, EVs utilizing FCs are power-driven by hydrogen with zero emissions since they only release warm air and water vapor. They generate electrical energy from fuels with chemical reactions simply operating as an electrochemical converter without combustion. They are much more effective than vehicles with straight internal combustion engines. In addition, FCs are availing to have higher power density and, thus, superior storage capacity than batteries [9]. The type of FCs can be classified into six groups and entitled depending on the employed electrolytes; (i) Alkaline (**AFC**), (ii) Molten carbonate (**MCFC**), (iii) Solid oxide (**SOFC**), (iv) Phosphoric acid (**PAFC**), (v) Direct methanol (**DMFC**), and (vi) Proton exchange membrane (**PEM**) FCs.

Consequently, the FC market is predicted to receive a rising demand due to the increasing use of fuel cell-powered vehicles from 2020 to 2025. The schematic diagram depicted in Figure 6.2 summarizes the FCs Market

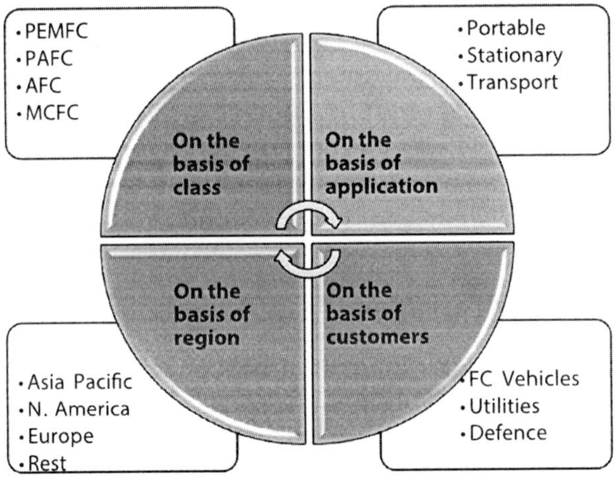

Figure 6.2 Categorization of the Fuel Cells Market based on type, region, customer, and use.

based on type, region, client, and use. PEMFC, PAFC, AFC and MEFC are the most common FCs for portable, stationary, and transport applications. N. America (including Canada and US), Europe, Asia Pacific (China, S. Korea and Japan) are the leading regions for FCs production and demand. The types of FCs are tabulated in Table 6.1 based on the properties, operation conditions, the most influencing brands in market, pros and cons.

6.1.3 Recent Developments in FC Vehicles (FCV) Market

FCVs market has been expanded tremendously in recent years. In 2018, Hydrogenics inaugurated a new provision in California, wherein it plans to integrate hydrogen FC systems into heavy-duty truck and bus platforms for the clients. Plug Power floated a new product for hydrogen-based FCs. Worldwide, approximately 13.000 FCVs were on the road in 2019, mainly in the Japan, China, USA, and Europe [10, 11].

The products by ProGen industry is leading original equipment manufacturers (OEMs) to approve sustainable FC power. ProGen 30kW engine offers EVs with prolonged runtimes with zero emissions. Ballard Power contracted with BEHALA (Berlin-based) to supply three FCveloCity 100 kW fuel cell modules in 2019. In 2020, The PEM and hydrogen FCs were used as energy supply for canal boats, named Elektra between Berlin and Hamburg and within Berlin [12]. SFC Energy engaged with Aurorahut (Finland based) to participate in the EFOY FC for a new environmentally friendly and noiseless igloo houseboats in 2019 [13]. Additionally, in

Table 6.1 Comparison of FC technologies.

Fuel cell	T_o range (°C)	Electrolyte	Charge carrier	Catalyst	Efficiency (%)	Benefits	Weaknesses	FC market
SOFC	650-1000	Yttria stabilized zirconia	O^{2-}	$LaMnO_3$/ $LaCoO_3$	55-65	Air as oxidant High efficiency	High operation temperature	Bloom Energy
MCFC	600-700	Li/K carbonate	CO_3^{2-}	Nickel	60-65	Air as oxidant High efficiency	High operation temperature	FuelCell Energy
PAFC	150-220	H_3PO_4	H^+	Pt	55	Insensitive to CO_2	Sensitive to CO Slow start	Doosan
AFC	50-90	KOH	OH^-	Pt	40-60	Quick start Room Temperature	require pure O_2 as oxidant	AFC Energy
DMFC	50-120	Polymer Membrane	H^+	Pt /Ru	30-40	Simple system Easy fuel storage	Cross over Limited lifetime	SFC Energy
PEMFC	60-200	Polymer Membrane	H^+	Pt	40-60	Quick start Room temperature Air as oxidant	Sensitive to CO Reactants need to be humidified	Ballard Power

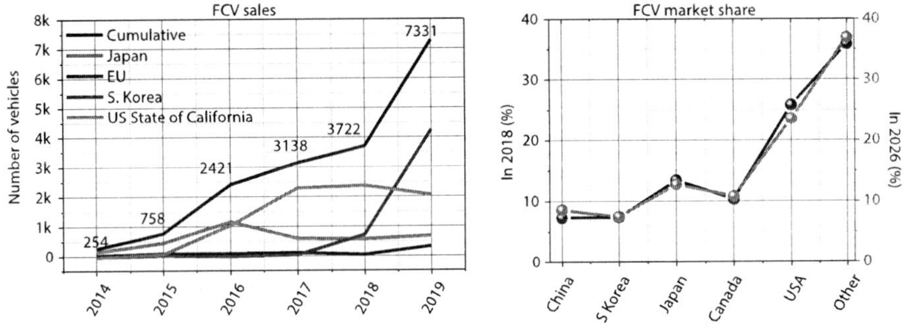

Figure 6.3 The sales of global hydrogen FCV yearly and cumulative between the years of 2014-19 (Created base on the data from Allied Market Research) [7], Share prediction of top five countries for FCV market between 2018-2026 (reproduced from the data of [14]).

2019, The California Fuel Cell Partnership has specified targets for 1.000 hydrogen refueling stations and 1.000.000 FCEVs by 2030. Eventually, The European Strategic Energy Technology (SET) Proposal has recognized FC and hydrogen technologies as critical technologies to achieve the target of the integrated European Energy and Climate Policy within 2020 and beyond [11].

Figure 6.3 indicates the sales of global hydrogen FCV yearly and cumulative from 2014 to 2019 and prediction for the top five countries for FCV market between 2018 and 2026. In 2020, Asia Pacific was the largest region, and North America was the second-largest one in the hydrogen fuel cell market. In July 2021, Hyzon Motors Inc. broadcasted a new commercial vehicle at lower cost and reduced weight (joined with lightweight composite materials) of a commercial vehicle powered by Hyzon's hydrogen fuel.

6.2 PEMFC Technology

6.2.1 PEMFC Working Principle and Components

In PEMFCs, during the process, hydrogen is fed into the anode, while air is fed into the cathode flow channels (as the oxidizer), leading to the electrochemical reactions given below;

$$\text{Anode: } 2H_2 \rightarrow 4H^+ + 4e^- \qquad \textit{Hydrogen oxidation (HOR)}$$

$$\text{Cathode: } O_2 + 4H^+ + 4e^- \rightarrow 2H_2O \qquad \textit{Oxygen reduction (ORR)}$$

$$\text{Overall: } 2H_2 + O_2 \rightarrow 2H_2O$$

Membrane Electrode Assembly (MEA) is the fundamental component of a PEMFC and it is composed of a polymer electrolyte membrane packed in between the cathode and anode electrodes. This polymeric membrane provides proton conduction, electronic insulation, and separation of gas reactants. The electrodes consist of the Catalyst Layer (CL), the Microporous Layer (MPL), and the Gas Diffusion Layer (GDL). The MEA is placed between two Bipolar Plates (BP), where gas flow channels (GFC) are located (Figure 6.4a-d). The HOR and ORR occur at the boundaries in the anode and cathode CLs, respectively. A GDL/MPL is placed between the CL and BP to conduct electron and heat, transport gas reactants, and enable water management.

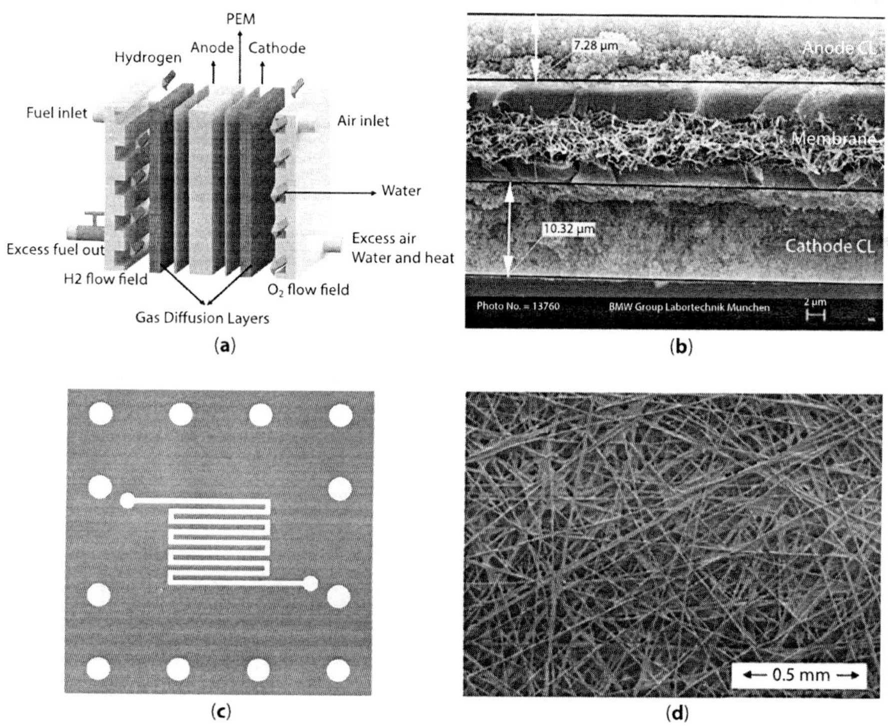

Figure 6.4 (a) Schematic representation of a typical PEMFC; HOR takes place at the anode, while the ORR occurs and at the cathode (b) Cross sectional SEM image [15], (c) Bipolar plate, (d) Scanning Electron Microscope (SEM) image of a carbon paper as a typical gas diffusion layer [16].

6.2.1.1 *Proton Exchange Membrane*

A polymer membrane (usually Nafion®) works as an electrolyte in the PEMFCs supplying proton conduction, electronic insulation, and gas separation (Figure 6.5) [17, 18]. Therefore, chemically inertness, high ionic conduction, and electron transport inhibition are much appreciated for a membrane material. The target for the proton conductivity of 0.1 $S.cm^{-1}$ at 120 °C and 50% relative humidity (RH) has been set by The United States Department of Energy (DOE) [19, 20]. In the early applications, a thick membrane like Nafion®117 (183-μm-thickness) was used for the mechanical strength and gas separation. Nowadays, due to the material advancements in mechanically strengthened membranes, thin membranes such as Core-Select™ are more popular.

Standard electrolyte membranes which operate below 100 °C are based on perfluorosulfonic acid (PFSA) polymers. Dupont's Nafion® is the most used membrane in this family. Nafion® holds the advantage of hydrophobic perfluorinated backbone that provides the chemical stability, whereas promoted water absorption can be achieved via hydrated clusters of hydrophilic sulfonated side chains [21]. Besides, it shows mechanical and chemical strength and admirable conductivity below 80 °C and in a high humid environment (0.13 $S.cm^{-1}$ at 75 °C of proton conductivity and

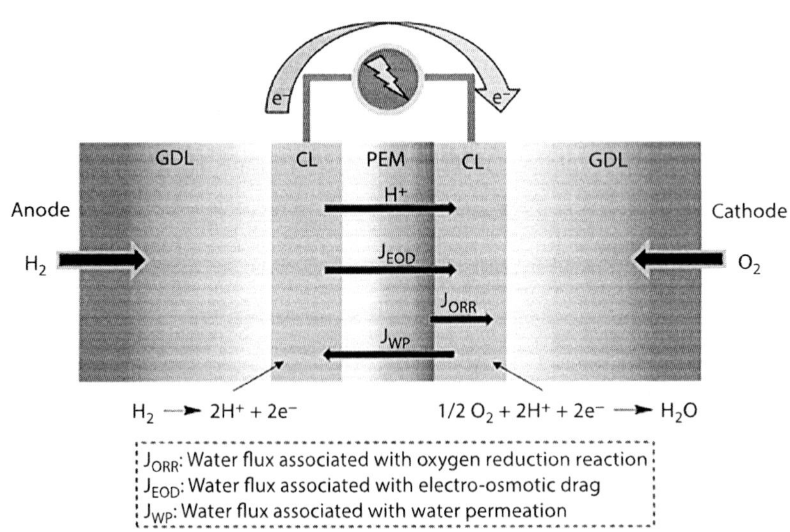

Figure 6.5 Transport of water within a PEMFC [18].

100% RH, durability above 60 000 h). Nevertheless, proton conductivity of Nafion® membrane drops dramatically above this temperature and restricts the application of Nafion® within the range of moderate temperatures (50–90 °C) [22]. Since higher temperatures morphologically favor the less interconnected ionic cluster formation, although there is no chemical change in the structure which was early predicted that this was a reason for poor proton conductivity. The properties of common fluorocarbon polymeric membranes used for PEMFC systems can be found in literature [23].

The significant drawbacks of the PFSA membranes such as high production cost, high gas permeability, and environmental incompatibility, have driven the research community to pursuit alternative membranes. For instance, it has been revealed that Dow Company and Solvay's membranes have higher proton conductivity, crystallinity, and glass transition temperature than that of Nafion®. In various studies, additives such as inorganic metal oxides (ZrO_2, TiO_2, $TiSiO_4$), clay, and silica into the matrix have been proposed to improve water retention and proton conductivity, as well as extend durability at higher temperatures [24]. Lately, cerium oxide additive into polytetrafluoroethylene matrix has been reported to increase the durability in PEMFC [25]. Additionally, 30 μm-thick "paper" membrane from sulfonation of cellulose nanofibers prepared with spray deposition [26]. A very recent work reported by Lim *et al.*, proposed the protonated phosphonic acids comprising tetrafluorostyrene-phosphonic acid and perfluorosulfonic acid polymers used in heavy duty FC and reached to power density of 780 $mW.cm^{-2}$ at 160 °C during 2500 h of operation and 700 thermal cycles from 40 to 160 °C [27]. In the study, perfluorosulfonic acid proton is transferred to the phosphonic acid to enhance the anhydrous proton conduction and durability of FC electrodes.

Sulfonated hydrocarbon polymers come into prominence as highly promising membranes owing to their outstanding properties such as excellent capacity for film formation, high water uptake capability at high temperatures and low humidity, as well as good thermal, mechanical, and chemical stability [18, 28–33]. Besides, the physicochemical properties can be easily tuned by functionalizing monomers or modifying the final polymers. In literature, the physicochemical and electrochemical properties of sulfonated polymer membranes having various degrees of sulfonation (DS) have been reported for FC applications [28, 34, 35]. For instance, Y.S. Kim *et al.* have explored the effect of degree of sulfonation on density, ion exchange capacity (IEC), water uptake and proton conductivity to build a correlation between structure-property and FC

Table 6.2 Comparative characteristics of some proton-conducting polysulfone composite membranes and FC performances.

Membrane	Membrane properties			FC performance	Year/ref.
	WA (%)	IEC (meq.g^{-1})	Ionic conductivity (S.cm^{-1})		
PSU/SPANI/Nb_2O_5	17.6	1.50	0.067	NA	2021/[37]
sPSU/SGO	39.0	1.32	0.00094	PD:82.6 mW.cm^{-2} @ 110 °C	2021/[38]
Crosslinked CDNs/SPPSU	134.0	1.67	0.0563@80°C	OCV:1.02V@100%RH	2020/[39]
sPSU/LDH	31.0	1.49	0.0137@120°C	PD: 204.5 mW.cm^{-2} @110°C	2020/[40]
SPSU/NIMs/GO	34.1	1.49	0.23@75°C	OCV:1.038 V; PD: 167.6 mW.cm^{-2} @60°C	2019/[41]
SPEESSA/sulfonic acid zeolite	29.1	3.19	0.124	OCV:0.91V; PD: 0.45W.cm^{-2} @1.1A.cm^{-2}	2019/[42]
PSF/MOF/Si	16.5	0.86	0.017@70°C	OCV:0.9V; PD: 40.8mW.cm^{-2}@160°C	2018/[43]

performance [28]. In another research, Sankir *et al.* showed the critical points for large-scale production of one of the critical starting materials for synthesizing disulfonated copolymers [29]. Additionally, depending on the disulfonation degrees, the membranes can be adopted for specific hydrogen generation processes (e.g. micro-fuel cells, on-board hydrogen generation) [33]. Lately, Zuo *et al.* has revealed a study on sulfonated microporous polymer membranes for PEMFC applications exceeding the maximum power output of Nafion 117, reaching up to 370 $mW.cm^{-2}$ [36].

In recent years, numerous studies on proton-conducting polysulfone-based composite membranes have been reported to adjust polymeric membranes for enhanced PEM properties. The proton-conducting polysulfone-based composite membranes in recent PEMFC studies have been tabulated in Table 6.2.

6.2.1.2 *Electrodes*

A PEMFC electrode encompasses the CL, GDL and a MPL components [44]. The electrochemical reactions take place in CLs. Membrane is sandwiched between two CLs which are comprised of a highly active carbon-supported Pt or alloyed Pt catalyst. This three-layered structure enables the transportation of all the reactant species such as proton conduction in the ionomer network, in procure of gaseous reactants, water removal in the pore, and electron conduction in the Pt-carbon networks [45, 46]. Pt is a noble metal with a high catalytic activity used in ORR reactions at the cathode. However, the high cost of Pt and sensitivity to CO poisoning at standard operating temperature are the major issues to overcome [47]. Moreover, DOE emphasizes the cost reduction of PEMFC, especially the catalyst cost which corresponds to approximately 45% of the total FC stack cost [48]. Based on DOE, the performance index of MEA with Pt loading should be less than 0.125 $g.kW^{-1}$, whereas power density should be greater than 1.0 $W.cm^{-2}$ by the year 2020 [49].

Various research studies have been reported based on increasing ORR activity by reducing the Pt loading (i.e. Toyota Mirai 2nd generation vehicles) and replacing it with more economical catalyst materials [50]. These reports have mainly focused on the efficacy of morphology, nanostructure, and composition of Pt and Pt-based metal alloy electrocatalyst [46]. Morphology based studies rely on increasing the specific/mass activity of the catalyst by designing the morphology architectures such as nano-wires, nano-crystals, nano-cages, etc. [48, 51, 52]. Introducing non-noble metals to make an alloy with Pt is an efficient route to enable high activity while reducing the cost of electrocatalysts. Similarly, PtNi/C alloy with a low amount of Pt (2.76 wt%) achieved a power density of 1.1 $W.cm^{-2}$ [53]. Recently, a highly-durable alloy catalyst by alloying Pt-Pd with transition metals (Cu, Ni or Co) in ternary compositions has displayed promising results for mass commercialization [54]. In another study, Pt_3Ni nano-frame catalysts have been reported to have an enhancement in mass activity compared to commercial Pt/C catalyst due to enlarged Pt surfaces [55]. Compared to cathode catalyst, fewer studies have been conducted for the anode side, since the HOR reaction at the anode is much faster and requires a low amount of Pt. Besides Pt catalyst, various bimetallic catalysts containing Pd, Co, Ir and Ru have been investigated [56]. For instance, a HOR catalyst prepared from exfoliated carbon nanotubes supported Pd_3Co has been reported to reach up to 327 $mW.cm^{-2}$ of power density at 60 °C. This value corresponds to 65% of the power density obtained from commercial Pt/C [57].

The role of GDL with MPL is to supply mechanical support for MEA, electron conduction, and passage pathways for reactants/products. It is usually 100-300 mm thick layer and made of conductive carbon-based materials such as carbon paper, carbon cloth, nonwoven cloth, etc. These materials are estimated to be the conventional choice for GDL due to the advantages of low cost, chemical stability, electrical conductivity, and mechanical strength [48, 58]. The GDL materials are usually treated with hydrophobic agents, like polytetrafluoroethylene (PTFE) to improve the water flooding capacity and supply effective gas channels. Laser perforation is a potential alternative for structural modification of GDL [59].

6.2.1.3 Bipolar Plate (BP)

In PEMFCs, BPs provide heat management, mechanical reinforcement, water management and ensures the uniform distribution of reactants on the GDL/CL surface [60]. Traditional BPs are made of graphite due to their resistance against corrosion and high electrical conductivity. However, it has large gas permeability, and its brittleness obstructs long-term use and mass production. Nevertheless, various BP materials have been investigated for commercial PEMFCs, including carbon composites, aluminum, stainless steel, and titanium [61].

Carbon composite materials for BPs consist of polymer binder, and conductive carbon fillers provide electron and heat conductivity and mechanical strength, and gas impermeability. There are several studies based on the addition of various fillers such as carbon black, carbon fibers, graphite particles, and carbon nanotubes to enhance the composite conductivity [62]. Polycarbonate (PC) filled with carbon nanotube (CNT), carbon fiber (CF), graphite (G), and their double and triple hybrids have also been investigated as BPs material [63].

The advantages of metallic BPs such as high electrical and thermal conductivity, low gas permeability, and high mechanical strength makes it stand out material for PEMFCs. However, the corrosion issue of the metals in the acidic environment should be resolved. Protective coating is a common way to improve the corrosion resistance of metallic BP [64]. For instance, metallic, carbon-based, and composite coatings are commonly applied on aluminum BPs. Metallic nitrides such as titanium nitride (TiN) offer good corrosion resistance and low interfacial contact resistance. For example, TiN/Ti double-layer coating has exhibited performance enhancement as compared to uncoated bipolar plates in the actual FC tests [65]. The corrosion resistance of stainless steel depends on the types of alloys. Specifically, the 316 and 316L stainless steel have been found to be more suitable for BPs [66]. Additionally, the multilayer CrC coating on 316L stainless steel has matched the DOE 2020 technical target in terms of corrosion resistance and electrical conductivity [67]. Some recent studies on BPs materials and coatings and the coating methods are compared and listed in Table 6.3.

Addition to materials, the topology of BPs is also a significant parameter for FC efficiency, since the flow field on channels of BPs provides effectual passage of reactants to reaction site and water management within the FC. The basic flow field designs used in PEMFCs are straight parallel, serpentine, pin-type, and interdigitated-type flow fields. Since the integration of PEMFCs into automotive applications, development and innovation in BPs have been pursued. Table 6.4 lists the recent studies on the flow field on channels.

Straight grooves structure is generally used in conventional cell flow field structures in FC stacks [75]. On the first generation Mirai, the FC stack with the conventional grooved flow field was replaced with a 3D fine-mesh cell flow field structure, while stainless steel was replaced with Ti in 2014 [48, 75]. In this way, the maximum power of the new stack has been increased from 90 to 114 kW (a per cell increase of 36%). Additionally, the change in the separator flow field material from stainless steel to Ti

Table 6.3 Several coating materials and their properties recently reported in the literature.

BP substrate	Coating material	Method	Results	Year/ Ref.
410 stainless steel	CrN	Cathodic Arc Evaporation (CAE)	Extending lifetime, corrosion resistance, lowering the cost	2022/ [68]
Ti-6Al-4V Titanium	Nitride	Powder immersion reaction assisted coating (PIRAC) method	As the nitrite content increased, the electrical conductivity increased	2020/ [69]
316L stainless steel	Amorphous carbon films	Direct current magnetron sputtering (DCMS)	Interfacial contact resistance decreased approximately two times	2020/ [70]
430 ferritic stainless steel	nanocrystalline β-Nb2N	Electric spark cutting machine	β-Nb_2N coating is decrease both the corrosion rate and Interfacial contact resistance of 430 FSS	2020/ [71]
303 stainless steel	Polyaniline/ Zn-Porphyrin composites	Spray gun	Excellent anti-corrosion activity and superior output power of the fuel cell.	2020/ [72]
Titanium	Gold	Electrodeposition	The total ohmic resistance of the fuel cell is reduced by 1.8 times	2019/ [73]
304 stainless steel	carbon nanotubes (CNTs)	Plasma-enhanced chemical vapor deposition (PECVD)	CNT coating makes it suitable for low-temperature PEM fuel cell applications.	2019/ [74]

Table 6.4 Different bipolar plate flow channels design and properties recently reported in the literature.

Flow channel images	Flow channel types	Material	Results	Year/Ref.
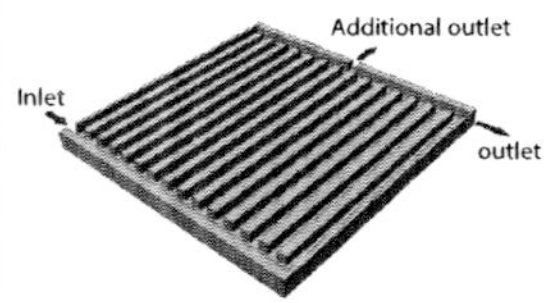	Parallel flow fields	NA	Regulated the oxygen concentration distribution in the flow area and increased the gas velocity in the flow channels	2021/[77]
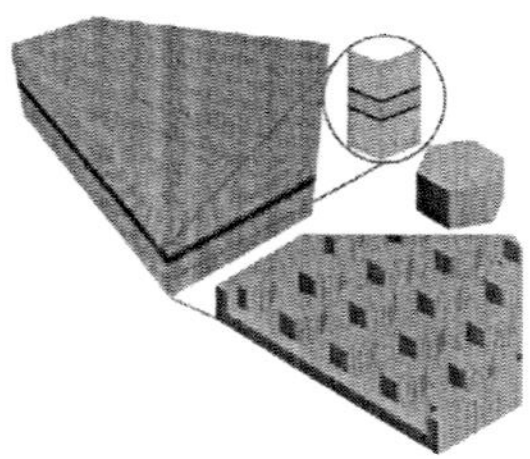	Honeycomb flow field	NA	Oxygen diffusion rate in the hexagonal pin (honeycomb) flow channel GDL is improved by 10-fold	2019/[78]

(*Continued*)

Table 6.4 Different bipolar plate flow channels design and properties recently reported in the literature. (*Continued*)

Flow channel images	Flow channel types	Material	Results	Year/Ref.
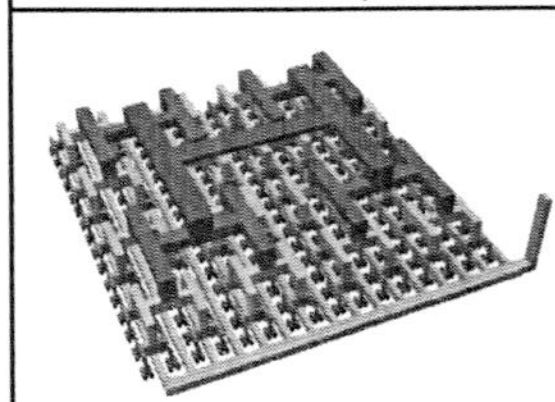	Lung-inspired, flow-fields	Aluminum plates	Lung-inspired flow-fields contributed for preventing backflow of fluid	2019/[79]
	Leaf bio-inspired flow field	*Lexan*™ resin	Improved water management due to branching	2015/[80]

(*Continued*)

Table 6.4 Different bipolar plate flow channels design and properties recently reported in the literature. (*Continued*)

Flow channel images	Flow channel types	Material	Results	Year/Ref.
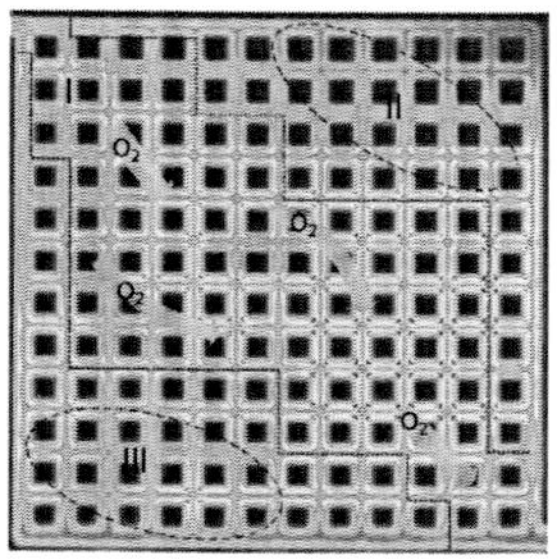	Square mesh flow field	Stainless steel plates	The square mesh flow field increased the catalytic activity by uniformly dispersing the gas and liquid.	2022/[81]

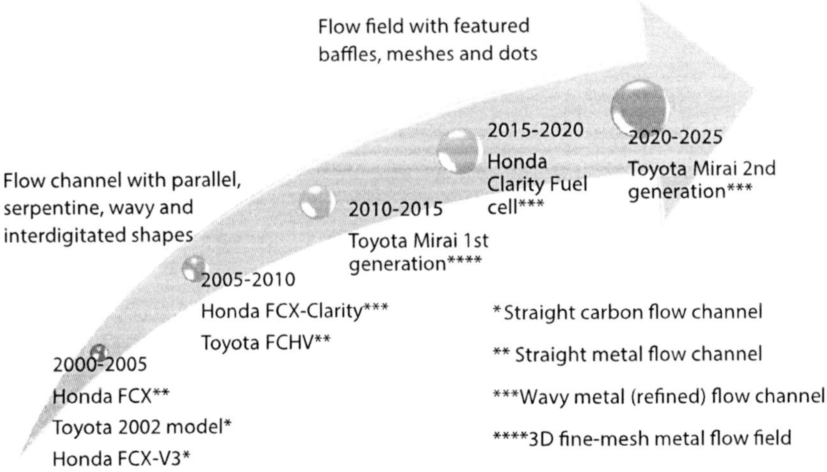

Figure 6.6 The timeline of the BPs developments in PEMFC EVs.

reduced cell weight by 39%. In 2020, on the second generation Mirai, 3D flow field was replaced with a two-dimensional (2D) wavy channel, since the previous one had fracture surface cracks after exposure to an acidic electrochemical environment [76]. Figure 6.6 demonstrates the innovations of BPs in electrical vehicles.

6.2.2 Fuel Cell Efficiency

It is not proper to calculate the efficiencies of FC based on Carnot cycles, where hot and cold reservoirs simultaneously produce work because the efficiency is highly dependent on the temperature differences between hot and cold reservoirs. There is no combustion, or the idea related to combustion happening in the fuel cells. Only the electrons are produced by consuming a chemical potential of reactants. FC effort is purely for producing the electrical energy, which is converted from chemical energy by electrochemical reactions. However, there are many attempts to better define efficiencies of fuel cells. For example, the thermodynamic efficiency can be defined as;

$$e = \left[\frac{\text{work}}{\Delta H}\right]$$

And the maximum available work is;

$$(dG)_{T,P} = dw$$

From the above equations, the efficiency becomes

$$e = \left[\frac{work}{\Delta H}\right] = \frac{\Delta G}{\Delta H}$$

This equation resembles how much enthalpy can be converted into practical work. There is always liquid water produced at the cathode; therefore, enthalpy and Gibbs free energy change for the high heating value of water, which is about 286 and 237 $kJ.mol^{-1}$, respectively. Hence, we can report about 83 % efficiency based on these thermodynamic parameters. This can be considered as maximum capacity of work can be extracted from the fuel cell. However, in reality, we have to define other efficiency concepts rather than thermodynamic efficiency. On the other hand, it is very efficient to operate a single cell at about 0.5-0.6 V, which is about half of the reversible cell potential. Then voltage efficiency can be described in the following ratio

$$e_V = \frac{E'}{E^0}$$

where E^0, E' is reversible cell voltages and the actual potential, which is used during fuel cell operates, respectively. Then, the fuel cell efficiency becomes about 50 percent when we consider the corrected $E^{0,c}$, about 1.1V. However, voltage efficiency is not a robust way to express efficiency, since the ratio itself is only dependent on the voltage. It is better to use the idea of voltage efficiency in terms of energy efficiency which is;

$$e_E = \frac{nFE^0 - nF\eta}{nFE^0} = 1 - \frac{\eta}{E^0}$$

where η is voltage loss to run fuel cell at higher current densities.

Therefore, if there is approximately 40 percent loss to operate the fuel cell at higher current densities at maximum power, the fuel cell efficiency is about 60 %.

The other efficiency term is called fuel efficiency. The ratio for reaction rate to fueling rate is;

$$e_F = \frac{r_R}{r_F} = \frac{ia_m / nF}{r_F} = \frac{i}{Fr_F},$$

where i and a_m are current and atomic mass, respectively. Since atomic mass (a_m) and n (equivalent electron) are equal, then fuel efficiency is;

$$e_F = \frac{i}{Fr_F}$$

For example, efficiencies for a fuel cell running with a current density of $1A.cm^{-2}$ at various active areas (5, 10 20 cm^2) and various flow rates (0.2, 0.5, and 1.0 $N.L.min^{-1}$) are theoretically in the range of 3.5 to 70 %. Therefore, by just changing the active surface are, fuel efficiency can be improved significantly. Moreover, fuel efficiency could be 100 percent with novel designs, which refuel unreacted hydrogen gas while fuel cell operates. As a result, based on the definition of efficiency, we can compare and contrast fuel cells with other systems producing DC electricity.

6.2.3 Challenges to Overcome for FCVs

The dissemination of the use of FC in transport sector depends on the growing FC technology. However, there are still many challenges for widespread commercialization in the FC industry. High cost, the durability of the units, and performance are essential factors that effect widespread use of FCVs. The high costs of the materials used in FC systems such as catalysts, membrane and bipolar plates, can be reduced by technological innovations that conduce to increase power density, enhance durability, and simplify system complexity. For instance, decreasing the amount of platinum used in the electro-catalyst layer would decrease the total cost of the PEMFCs [82]. The catalysts prepared by alloying Pt with low-cost metals and Pt-based nano catalysts help to decrease the amount of used Pt

[82, 83]. The safe storage for hydrogen and transport requires well designed infrastructure which is an extra cost and effort. Therefore, the scientific researches accelerated to achieve advanced FC components and develop hydrogen production and storage in more economical and efficient ways.

6.3 Hydrogen Storage for FCs and On-Demand Hydrogen Generation

6.3.1 Hydrogen Storage

Hydrogen storage technology is significantly important for the progress of hydrogen and FC technologies in transportation, stationary and portable power applications. The main task is to find an efficient technique to distribute hydrogen for use, since hydrogen has a very low density (0.09 kg.m^{-3}). Therefore, it is necessary to use physical (compression or liquefaction) or chemical methods to densify hydrogen [84]. Safety is another critical parameter for hydrogen storage. Additionally, the rate of kinetics must fit in the transportation applications. The rate of kinetics can be described as the hydrogen generation rate by the system upon demand and stop it when required. Besides, storing hydrogen inside a vehicle requires proper design and architecture [85]. In Figure 6.7, the hydrogen storage tank installed in a Mazda RX-8 Hydrogen car can be viewed.

The hydrogen storage technologies can be divided into two main groups as (i) physical-based and (ii) material-based methods [87]. The subsections

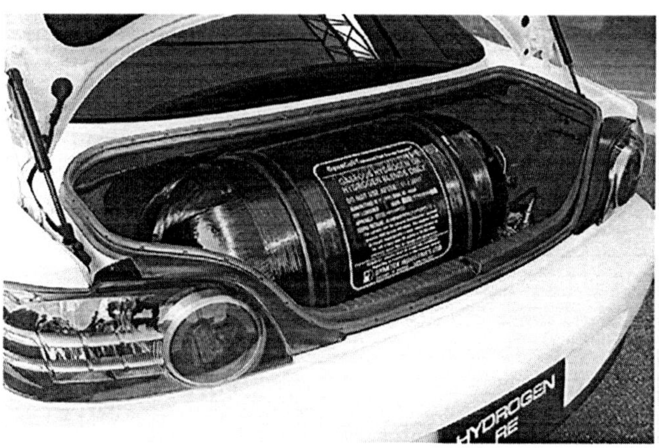

Figure 6.7 The hydrogen storage tank installed in a Mazda RX-8 Hydrogen car [86].

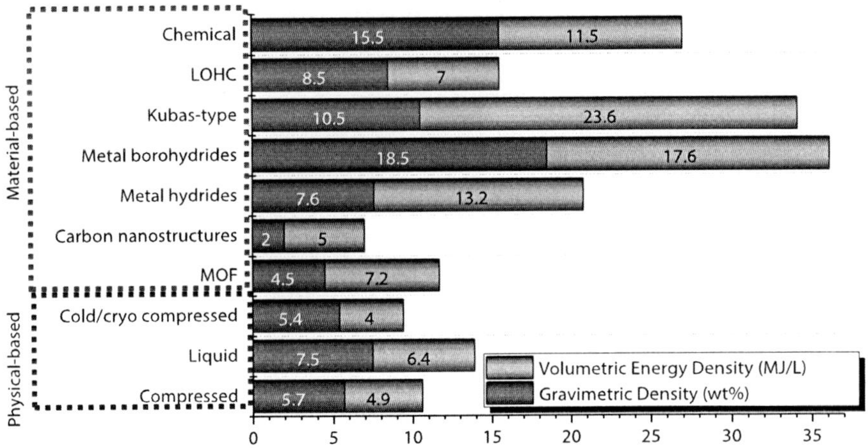

Figure 6.8 Hydrogen storage techniques with comparative volumetric (MJ/L) and gravimetric (w%) densities.

of the storage methods for hydrogen and their properties in terms of volumetric and gravimetric energy densities are summarized in Figure 6.8.

6.3.1.1 *Physical-Based Hydrogen Storage*

Hydrogen gas can be stored physically in three storage technologies; compressed gas, cold-cryo compressed, and liquid hydrogen. To date, compressed gas is the most confirmed hydrogen storage technology in the industry [87]. Honda Clarity and Toyota Mirai, both count on pressure vessels for on-board hydrogen storage [88]. Pressure vessels are classified based on their materials; Type I: all-metal, Type II: mostly metal-composite overwrap in the hoop direction, Type III: metal liner-full composite overwrap, and, Type IV: all composite constructions. For instance, Toyota Mirai uses carbon-fiber-reinforced tanks (Type IV). Based on The Toyota Motor Corporation, Mirai has 5.6 wt% of hydrogen tank capacity with 70 MPa of normal operating pressure [89].

It is required to store hydrogen in liquid form (LH2 storage) at cryogenic temperatures due to the low boiling point of hydrogen (−252.8°C) at ambient pressure. LH2 storage shows high potential for efficient hydrogen storage and transportation due to its high gravimetric and volumetric energy densities and hydrogen purity [90]. However, basic safety requirements for storage and transportation must be measured carefully. Liquid hydrogen tanks must be heavily insulated due to the low temperature which results in additional cost. Moreover, its thermodynamic feature makes mobility more difficult. Hydrogen storage as Cold/Cryo compression (CcH2) technology is regarded as a hybrid method of gas and liquid form of hydrogen. Therefore, the tank design should be suitable to endure the internal pressure of cryogenic fluid [91]. In 2011, BMW designated a prototype with CcH2 compression technology, using Type III pressure vessel, enhanced the gravimetric and power densities 50% more than that of LH2 compression [92].

6.3.1.2 *Material-Based Hydrogen Storage*

The researchers upon material-based storage has focused on sorbent materials, metal hydrides, and chemical hydrogen storage as a long-term solution for the issues of on-board electrical vehicles storage, as well as prospects on stationary and portable power applications [93]. Figure 6.9 demonstrates various hydrogen storage materials along with their hydrogen gravimetric capacity as a function of hydrogen release temperature explored by The Fuel Cell Technologies Office's (FCTO's) [94].

Adsorption-Based Materials

Carbon-based nanostructures such as carbon nanotubes (CNTs) and nanofibers as well as porous carbon are potential hydrogen adsorption materials

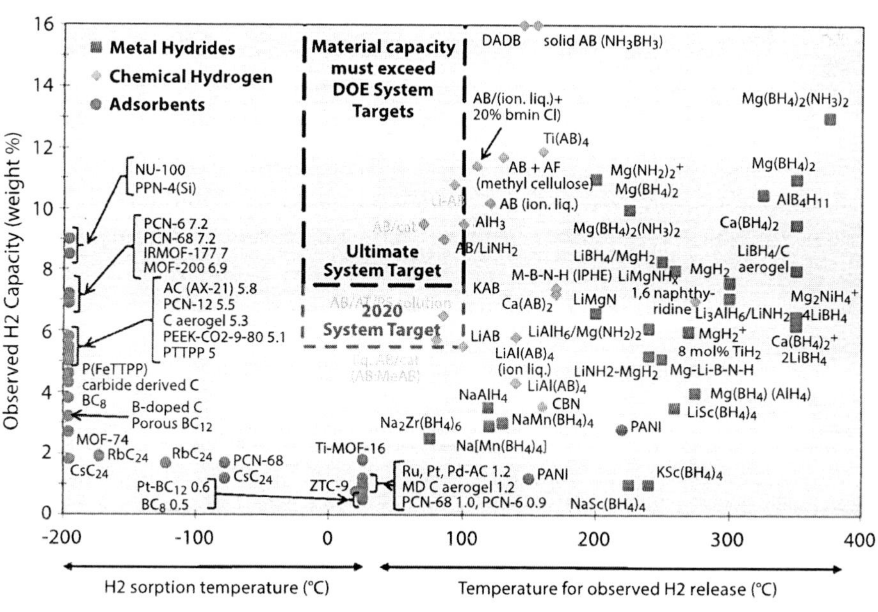

Figure 6.9 Hydrogen storage materials listed by FCTO [94].

owing to their high specific surface area. Especially, activated carbon is the most used carbon-based material under cryogenic conditions. In 1997, a research-based on single-walled carbon nanotubes had displayed a high hydrogen storage capacity (5 wt%–10 wt%) at the temperature of 273 K and pressure of 0.04 MPa [95]. Since the results have reached the hydrogen storage target announced by DOE, CNTs have been considered as an ideal hydrogen storage material. This study has been followed by several researches on CNTs, experimentally and theoretically, however, the outcomes are inconsistent. Consequently, there is still a debate on whether the CNTs meet the DOE's target or it is not [96].

In the early 2000s, a new group of highly porous, crystalline materials with large surface area (up to 4800 $m^2.g^{-1}$) named as Metal-Organic Frameworks (MOFs) have appeared for the researches on hydrogen storage materials. Over the past two decades, several studies have been reported on various types of MOFs for hydrogen adsorption. MOF-5 exhibited H_2 uptake at maximum of 1.3 wt% at 1 bar [97]. The results show that MOF-5 reached the DOE targets in 2020 for the cost and performance parameters of the automotive material-based hydrogen storage [98]. It is noted that the adsorbing performance of these structures strongly relies on their surface area and interaction of their active sites with hydrogen [99–101]. Also, the tunability of pore size depending on design methods is an excellent advantage for hydrogen adsorption [102]. For instance, MOF structure NU-100 has displayed to adsorb 14 wt% hydrogen at 100 bar pressure [103].

Absorption-Based Materials

Liquid organic hydrogen carriers (LOHCs): A variety of cycloalkanes, methyl cyclohexane-toluene, amino alcohols, and ammonia borane-based systems are classified as liquid organic hydrogen carriers (LOHCs) in hydrogen storage materials [104–106]. LOHCs systems where hydrogen forms a covalent bond with a liquid carrier enable the storage and transportation of hydrogen energy through hydrogenation and dehydrogenation processes. The biggest issue with LOHCs systems is the high cost and harsh operating conditions such as high temperature and pressure. Addition to economic aspects, toxicity is also an important issue for their mobile applications.

Metal hydrides: Metal hydrides are compounds of hydrogen with lightweight elements (Li, B, N, Mg and Al) and display extreme performance for hydrogen storage at low temperature and moderate pressures [107–109]. For instance, aluminum hydride, AlH_3 contains 10.1 wt% of hydrogen with a density of 1.48 H_2 $g.mL^{-1}$. Consequently, this material group has been extensively explored in literature [110, 111]. It can be considered as the safest hydrogen storage due to the endothermic hydrogen generation at relatively low operating temperatures. However, the low gravimetric capacity, slow kinetics and low reversibility are the main drawbacks for this type of hydrogen storage materials [112, 113]. Additionally, high temperature is necessary for the hydrogen desorption process, slow desorption kinetics, and high reactivity toward air and oxygen are the other disadvantages of metal hydrides as storage material [114, 115]. In order to alter the present drawbacks, the studies based on hydrogen storage using nanoscale metal hydrides have accelerated due to the advantages of enhanced reversibility, faster rates with nanointerfacial reaction, and altered heats of hydrogen absorption/desorption, as compared to bulk stage [111]. Another way to increase the efficiency of metal hydrides is the utilization of additive materials [116]. For instance, addition of V-based solid solution alloy has been reported to help for a decrease in dehydrogenation temperature and enhance the kinetics of both absorption and desorption reactions [116].

Complex hydrides and borohydrides: Light elements such as B, N, Al form complex metal hydrides (CMHs) with high hydrogen densities. The use of complex hydrides for hydrogen storage started with Ti catalyzed $NaAlH_4$. Later on, the successful outcomes resulted in an extension of the topic towards metal borohydrides ($LiBH_4$, $NaBH_4$, $Mg(NH_4)_2$, $Al(BH_4)_3$) [117]. In the group, $LiBH_4$ has high gravimetric hydrogen capacity (18.5 wt%) and volumetric energy density (43.4 $MJ.L^{-1}$). However, very high temperature (above 400°C) is required for hydrogen generation, owing to its thermodynamically stable nature. On the other hand, $NaBH_4$ is a promising material having high hydrogen storage capacity and it has lower cost. In order to stabilize the $NaBH_4$ solutions, alkali hydroxides are used in the aqueous solutions of $NaBH_4$. The exothermic and spontaneous reaction can be sped up with a proper catalyst. Thus, a great deal of effort has been made to develop catalyst materials for $NaBH_4$ hydrolysis [85, 118–123].

Kubas-type hydrogen storage materials: Transition metal-decorated nanostructures are called Kubas-type hydrogen storage materials. For instance, titanium decorated-polyacetylene has 12 wt% hydrogen storage capacity [124]. In brief, a metal catalyst is integrated into an adsorbent storage material for Kubas-type interaction. During the Kubas interaction, H_2 bound to a metal, and an electron is donated from a H_2 σ-bonding orbital to an empty d-orbital of the transition metal [125]. In this way, the bond between two hydrogen atoms are not broken, however, lengthen. This interaction results in an elongation of hydrogen bonds (up to 20% of its original bond length) as compared to free H_2. Thus, a better interaction between hydrogen molecules and metal catalyst are provided via Kubas interaction [126–129]. In literature, a porous Kubas manganese hydride (KMH-1) has displayed hydrogen storage up to 10.5wt% (120 bar, room temperature) with reversible adsorption [127]. The computational studies and inelastic neutron scattering results revealed the high gravimetric hydrogen uptake to Kubas binding as the principal mechanism.

6.3.2 On-Board Hydrogen Generation

As described above, a vast diversity of materials is available for hydrogen storage technologies. However, each one of them has its own limitations to overcome for the further improvement of FCEVs [130]. Therefore, on-board hydrogen generation has been intensely studied for use in PEMFC. In the early applications, this process was conducted with alcohols (methanol, ethanol, dimethyl ether) and hydrocarbons (diesel, methane, natural gas, gasoline, and liquefied petroleum gas) within the systems requiring minimal start time, fast response, lightweight and robust components [131, 132].

A chemical reaction between aluminum powder and water that yields 1 kg of hydrogen per 9 kg of aluminum has been proposed for on-board hydrogen generation for aircraft [133]. In another study, hierarchical bulk nanoporous aluminum has been used with pure water to generate hydrogen for on-board applications [134]. Chemical hydrides are also suitable for on-board hydrogen generation for PEMFC applications [130, 135, 136]. In Figure 6.10, a setup containing a PEMFC system and a hydrogen generator filled with $NaBH_4$ solution is described for FC-powered small unmanned aircraft [135].

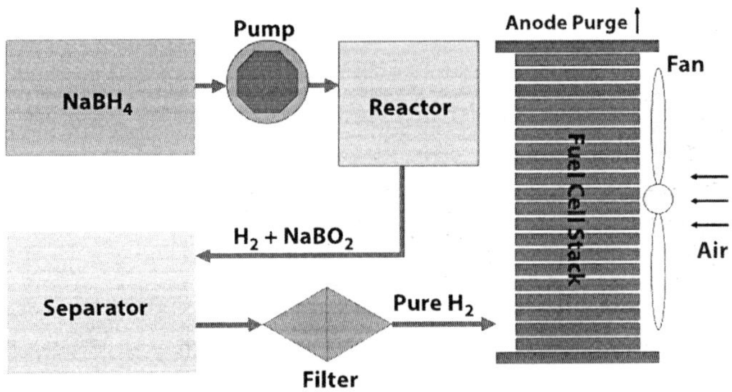

Figure 6.10 Schematic representation of $NaBH_4$ hydrogen generator attached to FC system [135].

6.3.3 Are the FCs Considered to be 100% Green?

Since the reaction of hydrogen with oxygen releases only water as tailpipe emission in an FCV, the process is considered as green. However, the production methods of hydrogen that is used as a fuel in FCV also matter in environmental concerns, because hydrogen is produced from a range of resources including fossil fuels, nuclear energy, biomass and renewable energy sources. The primary process to produce hydrogen, which is called "gray" hydrogen, is the reforming of natural gas involving methane and steam. The process produces a smaller amount of carbon emissions than "black" or "brown" hydrogen. Black or brown hydrogen is the most environmentally damaging as black or brown coal is used in the hydrogen-production process. The "blue" hydrogen is produced from natural gas with carbon capture. It means that the 10-20% carbon emitted from steam reforming is captured and stored underground through industrial carbon capture and storage. "Turquoise" hydrogen is produced from a process called methane pyrolysis, which generates solid carbon. At present, this type of hydrogen is still at the pilot stage. "Pink" hydrogen is generated through the electrolysis of water powered by nuclear energy rather than renewables. The last group is the "green" hydrogen (renewable) which is produced from the electrolysis of water driven by renewable energy sources. Today, green hydrogen accounts for about 0.1% of overall hydrogen production, but it is projected to increase as the funds on advanced technologies to utilize renewable energy continues to rise. The major groups of hydrogen production are given in Figure 6.11.

Since the production of blue and grey hydrogen involves natural gas, CO_2 emission cannot be neglected [138]. Thus, our ultimate goal must be to increase the prevalence of green hydrogen worldwide to reduce

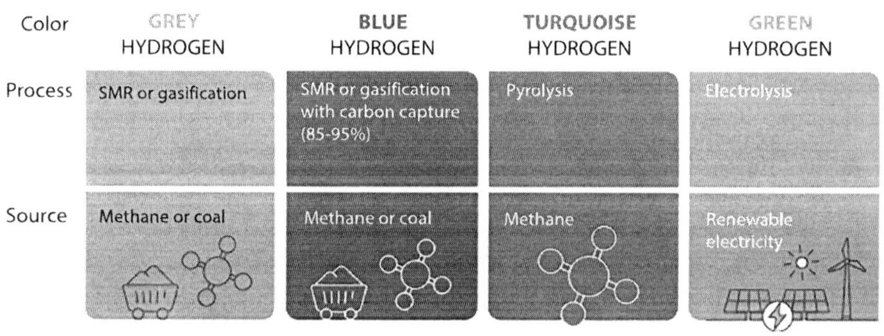

Figure 6.11 Classification of hydrogen (International Renewable Energy Agency, IRENA, 2020) [137].

our carbon footprint [139–141]. In this regard, green hydrogen-production methods are encouraged by the governments. For instance, the state of California demands that 30% of H_2 supplied for transportation sector should be obtained from the renewables such as biomass material, wind, and solar.

Consequently, scientific research has focused on advanced technologies [118] and nanomaterials to be used as effective (photo)catalyst materials for hydrogen production from renewables [142–155] and chemical hydrides [33, 119, 121–123, 136, 156, 157] to reduce the main issues of hydrogen, such as high-cost of production methods, safety in storage and logistics. Electrolysis of water into hydrogen and oxygen molecules using renewables such as wind and solar energy offers a promising footpath for sustainable hydrogen supply for FCEV. Based on the study conducted by Landman *et al.*, solar-to-hydrogen (STH) conversion efficiency has been reported as 7.5%, which can be improved to 10% of standard value for commercialization [150]. After a short time, a photoelectrochemical (PEC) cell combining III–V-based photo absorber and IrRuOx–Pt-based electrocatalysts has been reported to reach STH above 15%, verging on one more step towards commercial PEC device [158]. Today, the applicability

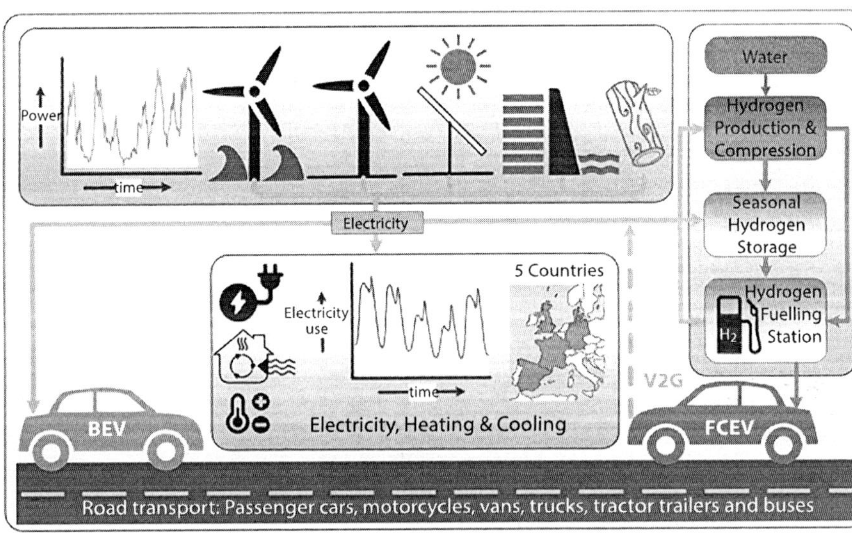

Figure 6.12 The proposed design for 100% renewable system that is applied to the national electricity, heating, cooling and transport systems of five European countries (Denmark, Germany, Great Britain, France and Spain). FCEVs to grid (FCEV and V2G), electrolyzers and hydrogen storage provide all of the necessary balancing requirements [159].

of the concept in Europe is under consideration for the European countries such as Denmark, Germany, Great Britain, France and Spain. Figure 6.12 presents the conceptional pathway using renewables to produce electricity and hydrogen for road transport in these five European countries [159].

6.4 FCs and Automotive Applications

Today, vehicle systems in the transport industry are categorized into three groups, as demonstrated in Figure 6.13. These groups are; (i) Internal combustion engine vehicles (ICEV), (ii) Hybrid electric vehicles (HEV), and (iii) electric vehicles (EV) which utilize ultra-capacitors and fuel cells for energy supply [160].

The Electrovan, which was designed in 1966 by General Motors (GM), was the first example for FCEV. This vehicle had a power source made of an AFC and two cryogenic tank containers for liquid hydrogen and oxygen [161, 162]. With the technology development, an extensive range of FC in the market, including low and high-temperature FC have been produced. Today, PEMFCs are the most favorable type for automotive applications owing to several reasons which led FCEV technology to be transferred from AFC to PEMFC in electric vehicles.

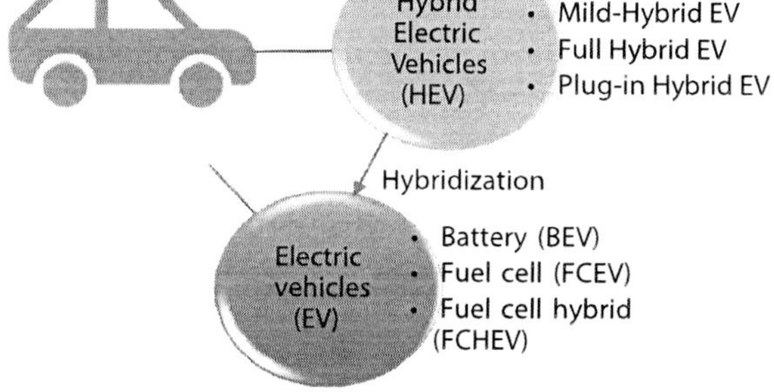

Figure 6.13 Classification of vehicles in transport industry (Reproduced from Ref. [160]).

Table 6.5 Commercialized FCEVs and their configurations [58, 175–184].

FCEVs	Stack power (kW)	Volumetric power density (kW/L)	Battery capacity (kWh)	Fuel tank capacity (kg)	Fuel pressure (MPa)	Range (mile/km)	Cold start (°C)
Hyundai NEXO	95	3.1	1.56	6.33	70	380/612	-30
Toyota Mirai I	114	3.1	1.6	5.00	70	312/502	-30
Toyota Mirai II	128	4.4	1.24	5.50	70	402/426	-30
Hyundai Tucson 2016	100	1.65	0.95	5.64	70	265/426	-20
Saic Maxus FCV80	115	3.1	-	6.00	35	312/502	-30
Honda FCX 2017	103	3.12	1.7	5.46	70	366/589	-30

The 3D fine-mesh flow channel was established for the first-generation MIRAI in 2014 (*). Later, 2D structure for the bipolar plates was used for the second-generation MIRAI in 2020 (**), possible owing to the high cost.

6.4.1 PEMFC Systems in Automobiles

PEMFC technology is the most favorable system for automobiles owing to their high-power density, low corrosion, lower operation temperature (60-80°C) and cold start capability [161, 163]. Additionally, they have the ability for operating under start-stop vehicle driving situations [164–167]. Thus, PEMFC technology is thought to have a great potential to reduce the energy challenges that the world is facing with, such as pollutant emissions from the transport sector, which is responsible for nearly 17% of the yearly global greenhouse gas emissions [168]. However, the transition in the transport industry from Internal Combustion Engine (ICE) vehicles to all EVs has been a long ride from the concept to marketing. Consequently, it is not a coincidence that the number of researches has constantly increased on the integration of PEMFC in electrical vehicles in the last few years [6, 19, 45, 160, 162, 169].

Initially, vehicles with slow speeds such as trams, submarines, buses, forklifts, were more suitable for FC applications. However, the current technological advancements have led to the utilization of fuel cells in high-speed vehicles. Modification of powertrain was an essential development for fuel cells [170–172]. Since 90s, the automobile sector including DaimlerChrysler, General Motors and Toyota has responded by devoting the researches on PEMFC. Ballard, meanwhile, was performing research and supplying PEMFC units for Ford and Daimler. In 2012, Hyundai was announced to be the first car manufacturer to commercialize FCEV [173].

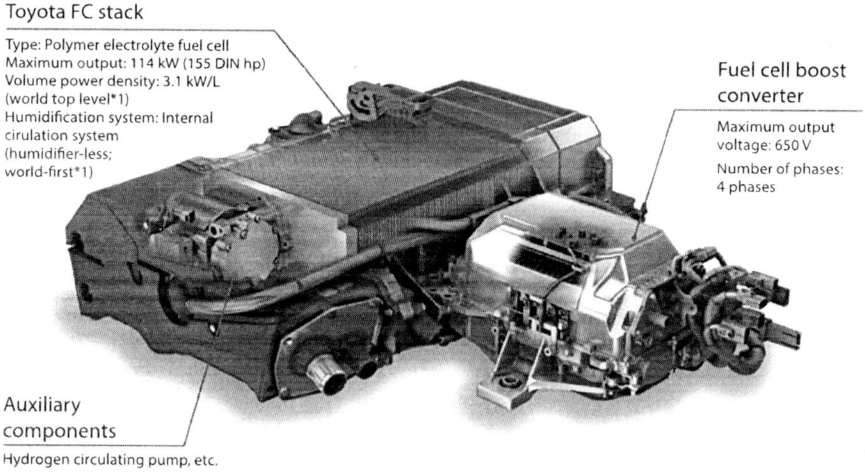

Figure 6.14 FC stack with BOP components in Toyota Mirai [58, 185].

Table 6.6 FCEBs available in Europe and US [58, 186].

FCEB	FC system	Max stack power (kW)	Battery capacity (kWh)	H_2 storage capacity (kg)	Equivalent gasoline/ diesel* (kg)	Range (mile)/(km)
ACT ZEBA	UTC Power	120	EnerDel/17.4	40	109.09/111.63	204/328
SL AFCB	Ballard	150	A123/11	50	136.36/139.53	260/418
UC Irvine AFCB	Ballard	150	A123/11	50	136.36/139.53	244/393
A330 FC	Ballard FCveloCity-HD85	85	24 or 36	38	103.64/106.05	220/354-250/402
Businova	Symbio H2Motiv	30	132	28	76.36/78.14	190/306
Streetdeck FCEV (double-decker)	Ballard FCveloCity-HD85	85	48	30	81.82/83.72	200/322-265/426 with increasing H_2 storage
H_2 City Gold	Toyota	60	29-44	37.5	102.27/104.65	250/402
Urbino 12 hydrogen	Ballard FCmove-HD	70	30	37.5	102.27/104.65	220/354

*Equivalent thermal energy. Estimate based on the lower heating values (LHV) of gasoline (44 $MJ.kg^{-1}$) and diesel (43 $MJ.kg^{-1}$), and hydrogen energy density (120 $MJ.kg^{-1}$).

This action was followed by Toyota Motor Corporation in 2014. Indeed, PEMFC has been commercialized with Mirai FCV, which was launched as a trademark by Toyota in 2017. H_2 is converted to power in the fuel-cell stack, producing 128 kW compared to the antecedent Mirai's 114 kW (Mirai I) [174]. Today, the common FCVs developed in the field are Honda FCX, Toyota Mirai I and II, and Hyundai Tucson FCV, Saic Maxus FCV80. The properties of these vehicles are tabulated in Table 6.5 along with their FC configurations.

FC stack with the balance of plants (BOP) components in Toyota Mirai is given in Figure 6.14. Besides FCEV, city buses powered by FC systems serve in many regions in the US and EU. For instance, there are more than 20 PEMFC powered buses are operating in four states. Table 6.6 displays the FCBs operating in US and EU along with FC configurations (Figure 6.15).

Figure 6.15 FC buses (FCBs) in California: Zero Emission Bay Area Demonstration Group led by Alameda-Contra Costa Transit District American FCB Project at SunLine Transit Agency, and American FCB Project at University of California, Irvine (UCI)) [186].

After the developments in FCEV and busses, fuel cell hybrid trains are under rapid progress [187]. For instance, Alstom's first fuel cell hybrid trains started their commercial use in Germany in 2017, while it completed its three months of trial in 2020 in Austria. Up to date, forty-one FC trains of Alstom have started to operate [188]. Siemens-Mireo tested a hydrogen-powered FC train in Bavaria and Baden-Württemberg in 2021. It is planned that this train will enter passenger service by January 2024 [189]. In addition, Ballard Power System and RWTH Aachen University have joint on a project to investigate the interaction between fuel cell systems and batteries [190].

Summary and Concluding Remarks

In this chapter, the global green energy requirements and the footsteps of the various governments for replacing fossil fuel with eco-friendly, sustainable, and more economical alternatives have been outlined. FCs has special place in the green energy consumption progress, especially in the transport sector. Among them, PEM fuel cells are most widely used in transportation due to various well-suited properties with electrical vehicle requirements. In the last decades, various vital developments have occurred in FC industry regarding the transport sector. Thus, the technological developments in the PEMFC market and current status of the FCEVs is required to be thoroughly reviewed. Hydrogen production, storage, and transportation are still crucial factors for FCEV developments. In this regard, this chapter includes hydrogen production and storage technologies. The progress in green hydrogen production methods using renewables and on-board hydrogen generation from hydrogen-rich chemicals have also been discussed for future applications.

References

1. Kirubakaran, A., Jain, S., Nema, R.K., A review on fuel cell technologies and power electronic interface. *Renew. Sustain. Energy Rev.*, 13, 9, 2430–2440, 2009, doi: 10.1016/j.rser.2009.04.004.
2. Kempa, K., Moslener, U., Schenker, O., The cost of debt of renewable and non-renewable energy firms. *Nat. Energy*, 6, 2, 135–142, 2021, doi: 10.1038/s41560-020-00745-x.
3. Taylor, M., *Energy subsidies: Evolution in the Global Energy Transformation to 2050.* International Renewable Energy Agency, Abu Dhabi, 2020.

4. IRENA International Renewable Energy, NDCs and Renewable, Energy Targets in and 2021, Are we on the right path to a climate-safe future?, pp. 12, Abu Dhabi, 2021.
5. van Renssen, S., The hydrogen solution? *Nat. Clim. Chang.*, 10, 9, 799–801, 2020, doi: 10.1038/s41558-020-0891-0.
6. Stern, A.G., A new sustainable hydrogen clean energy paradigm. *Int. J. Hydrogen Energy*, 43, 9, 4244–4255, 2018, doi: 10.1016/j.ijhydene.2017.12.180.
7. Khan, U., Yamamoto, T., Sato, H., An insight into potential early adopters of hydrogen fuel-cell vehicles in Japan. *Int. J. Hydrogen Energy*, 46, 18, 10589–10607, 2021, doi: 10.1016/j.ijhydene.2020.12.173.
8. Xue, M., Le Lin, B., Tsunemi, K., Emission implications of electric vehicles in Japan considering energy structure transition and penetration uncertainty. *J. Clean. Prod.*, 280, 124402, 2021, doi: 10.1016/j.jclepro.2020.124402
9. Neațu, Ș., *et al.*, Recent progress in electrocatalysts and electrodes for portable fuel cells. *J. Mater. Chem. A*, 9, 32, 17065–17128, 2021, doi: 10.1039/d1ta03644k.
10. Ajanovic, A. and Haas, R., Prospects and impediments for hydrogen and fuel cell vehicles in the transport sector. *Int. J. Hydrogen Energy*, 46, 16, 10049–10058, 2020, doi: 10.1016/j.ijhydene.2020.03.122.
11. IEA. Survey on the number of fuel cell vehicles, hydrogen refueling stations and target. *Adv. Fuel Cells IEA Technol. Collab. Program.*, pp. 1–7, 2019.
12. EST-Floattech. "ELEKTRA": World's first emission free hydrogen push boat, 2022. https://www.est-floattech.com/elektra-first-hydrogen-canal-tug (accessed Mar. 14, 2022)
13. SFC Energy AG. SFC Energy promotes soft tourism and wins follow-up order for EFOY fuel cells for AuroraHuts houseboat igloos, 2020. https://www.sfc.com/en/follow-up-order-for-efoy-fuel-cells-for-aurorahuts-houseboat-igloos/ (accessed Mar. 14, 2022)
14. European Hydrogen Association. Hydrogen and fuel cells. *EU Science Hub*, 2021. https://ec.europa.eu/jrc/en/research-topic/hydrogen-and-fuel-cells (accessed Mar. 14, 2022).
15. Obermaier, M., Bandarenka, A.S., Lohri-Tymozhynsky, C., A comprehensive physical impedance model of polymer electrolyte fuel cell cathodes in oxygen-free atmosphere. *Sci. Rep.*, 8, 1, 1–9, 2018, doi: 10.1038/s41598-018-23071-5
16. Wang, Y. and Gundevia, M., Measurement of thermal conductivity and heat pipe effect in hydrophilic and hydrophobic carbon papers. *Int. J. Heat Mass Transf.*, 60, 134–142, 2013, doi: 10.1016/j.ijheatmasstransfer.2012.12.016.
17. Mauritz, K.A. and Moore, R.B., State of understanding of Nafion. *Chem. Rev.*, 104, 10, 4535–4585, 2004, doi: 10.1021/cr0207123.
18. Adamski, M., Peressin, N., Holdcroft, S., Adamski, M., On the evolution of sulfonated polyphenylenes as proton exchange membranes for fuel cells. *Mater. Adv.*, 2, 4966–5005, 2021, doi: 10.1039/d1ma00511a.

19. Christmann, K., Friedrich, K.A., Zamel, N., Activation mechanisms in the catalyst coated membrane of PEM fuel cells. *Prog. Energy Combust. Sci.*, 85, 100924, 2021, doi: 10.1016/j.pecs.2021.100924.
20. Kim, D.J., Jo, M.J., Nam, S.Y., A review of polymer-nanocomposite electrolyte membranes for fuel cell application. *J. Ind. Eng. Chem.*, 21, 36–52, 2015. doi: 10.1016/j.jiec.2014.04.030.
21. Haile, S.M., Fuel cell materials and components. *Acta Mater.*, 51, 19, 5981–6000, 2003, doi: 10.1016/j.actamat.2003.08.004
22. Casciola, M., Alberti, G., Sganappa, M., Narducci, R., On the decay of Nafion proton conductivity at high temperature and relative humidity. *J. Power Sources*, 162, 1, 141–145, 2006, doi: 10.1016/j.jpowsour.2006.06.023
23. Kim, D.J., Jo, M.J., Nam, S.Y., A review of polymer-nanocomposite electrolyte membranes for fuel cell application. *J. Ind. Eng. Chem.*, 21, 36–52, 2015. doi: 10.1016/j.jiec.2014.04.030
24. Ain, N. and Harun, M., A review of alternative polymer electrolyte membrane for fuel cell application based on sulfonated poly (ether ether ketone). *Int. J. Energy Res.*, 45, 14, 19671–19708, 2021. https://doi.org/10.1002/er.7048.
25. Yoon, K.R., Lee, K.A., Jo, S., Yook, S.H., Lee, K.Y., Kim, I., Kim, J.Y., Mussel-inspired polydopamine-treated reinforced composite membranes with self-supported CeO_x radical scavengers for highly stable PEM fuel cells. *Adv. Funct. Mater.*, 29, 3, 1806929, 1–11, 2019. https://doi.org/10.1002/adfm.201806929.
26. Bayer, T., Vaughan, B., Br, C., Selyanchyn, R., Fujikawa, S., Sasaki, K., Matthew, S., Spray deposition of sulfonated cellulose nanofibers as electrolyte membranes in fuel cells. *Cellulose*, 28, 1355–1367, 2021. https://doi.org/10.1007/s10570-020-03593-w.
27. Lim, K.H., Lee, A.S., Atanasov, V., Kerres, J., Park, E.J., Adhikari, S., Maurya, S., Manriquez, L.D., Jung, J., Fujimoto, C., Matanovic, I., Jankovic, J., Hu, Z., Jia, H., Kim, Y.S., Protonated phosphonic acid electrodes for high power heavy-duty vehicle fuel cells. *Nat. Energy*, 7, 248–259, 2022. https://doi.org/10.1038/s41560-021-00971-x.
28. Kim, Y.S., Einsla, B., Sankir, M., Harrison, W., Pivovar, B.S., Structure-property-performance relationships of sulfonated poly(arylene ether sulfone)s as a polymer electrolyte for fuel cell applications. *Polymer (Guildf.)*, 47, 4026–4035, 2006. https://doi.org/10.1016/j.polymer.2006.02.032.
29. Sankir, M., Bhanu, V.A., Harrison, W.L., Ghassemi, H., Wiles, K.B., Glass, T.E., Brink, A.E., Brink, M.H., McGrath, J.E., Synthesis and characterization of 3,3′-Disulfonated-4,4′-dichlorodiphenyl sulfone (SDCDPS) monomer for proton exchange membranes (PEM) in fuel cell applications. *J. Appl. Polym. Sci.*, 100, 4595–4602, 2006. https://doi.org/10.1002/app.22803.
30. Mohammadi, M. and Mehdipour-ataei, S., Durable sulfonated partially fluorinated polysulfones as membrane for PEM fuel cell. *Renew. Energy*, 158, 421–430, 2020. https://doi.org/10.1016/j.renene.2020.05.124.

31. Han, X., Pang, J., Liu, D., Zhang, H., Jiang, Z., Novel branched sulfonated poly (arylene ether)s based on carbazole derivative for proton exchange membrane. *Int. J. Hydrogen Energy*, 45, 4644–4652, 2020. https://doi.org/10.1016/j.ijhydene.2019.12.120.
32. Zou, Y., Yang, M., Liu, G., Xu, C., Sulfonated poly (fluorenyl ether ketone nitrile) membranes used for high temperature PEM fuel cell. *Heliyon*, 6, e04855, 2020. https://doi.org/10.1016/j.heliyon.2020.e04855.
33. Sankir, M., Semiz, L., Sankir, N.D., Catalyst free hydrogen generation from directly disulfonated poly(arylene ether sulfone) copolymer membranes. *J. Memb. Sci.*, 496, 318–324, 2015. https://doi.org/10.1016/j.memsci.2015.07.059.
34. Kang, H. and Lee, D., Improving the durability and performance of sulfonated poly(arylene ether)s by introducing 9, 10-Dihydro-9-oxa-10-phosphaphenanthrene 10-oxide structure for fuel cell application, 2021. https://doi.org/10.1021/acsomega.1c04205.
35. Miyake, J., Kusakabe, M., Tsutsumida, A., Miyatake, K., Remarkable reinforcement effect in sulfonated aromatic polymers as fuel cell membrane. *ACS Appl. Energy Mater.*, 1, 1233–1238, 2018. https://doi.org/10.1021/acsaem.7b00349.
36. Zuo, P., Li, Y., Wang, A., Tan, R., Liu, Y., Liang, X., Sheng, F., Tang, G., Ge, L., Wu, L., Song, Q., McKeown, N.B., Yang, Z., Xu, T., Sulfonated microporous polymer membranes with fast and selective ion transport for electrochemical energy conversion and storage. *Angew. Chemie - Int. Ed.*, 59, 9564–9573, 2020. https://doi.org/10.1002/anie.202000012.
37. Maria Mahimai, B., Kulasekaran, P., Deivanayagam, P., Novel polysulfone/sulfonated polyaniline/niobium pentoxide polymer blend nanocomposite membranes for fuel cell applications. *J. Appl. Polym. Sci.*, 138, 1–10, 2021. https://doi.org/10.1002/app.51207
38. Simari, C., Lufrano, E., Brunetti, A., Barbieri, G., Nicotera, I., Polysulfone and organo-modified graphene oxide for new hybrid proton exchange membranes: A green alternative for high-efficiency PEMFCs. *Electrochim. Acta.*, 380, 138214, 2021. https://doi.org/10.1016/j.electacta.2021.138214.
39. Mohamad Nor, N.A., Nakao, H., Jaafar, J., Kim, J.D., Crosslinked carbon nanodots with highly sulfonated polyphenylsulfone as proton exchange membrane for fuel cell applications. *Int. J. Hydrogen Energy*, 45, 9979–9988, 2020. https://doi.org/10.1016/j.ijhydene.2020.01.142.
40. Simari, C., Lufrano, E., Brunetti, A., Barbieri, G., Nicotera, I., Highly-performing and low-cost nanostructured membranes based on polysulfone and layered doubled hydroxide for high-temperature proton exchange membrane fuel cells. *J. Power Sources*, 471, 228440, 2020. https://doi.org/10.1016/j.jpowsour.2020.228440.
41. Li, J., Wu, H., Cao, L., He, X., Shi, B., Li, Y., Xu, M., Jiang, Z., Enhanced proton conductivity of sulfonated polysulfone membranes under low humidity

via the incorporation of multifunctional graphene oxide. *ACS Appl. Nano Mater.*, 2, 4734–4743, 2019. https://doi.org/10.1021/acsanm.9b00446.

42. Munavalli, B.B. and Kariduraganavar, M.Y., Development of novel sulfonic acid functionalized zeolites incorporated composite proton exchange membranes for fuel cell application. *Electrochim. Acta.*, 296, 294–307, 2019. https://doi.org/10.1016/j.electacta.2018.11.056.
43. Ahmadian-Alam, L. and Mahdavi, H., A novel polysulfone-based ternary nanocomposite membrane consisting of metal-organic framework and silica nanoparticles: As proton exchange membrane for polymer electrolyte fuel cells. *Renew. Energy*, 126, 630–639, 2018. https://doi.org/10.1016/j.renene.2018.03.075.
44. Zhang, J., *PEM Fuel Cell Electrocatalysts and Catalyst Layers: Fundamentals and Applications*. Springer, London, 2008. doi: 10.1007/978-1-84800-936-3.
45. Ren, P., Pei, P., Li, Y., Wu, Z., Chen, D., Huang, S., Degradation mechanisms of proton exchange membrane fuel cell under typical automotive operating conditions. *Prog. Energy Combust. Sci.*, 80, 2020. https://doi.org/10.1016/j.pecs.2020.100859.
46. Borup, R.L., Kusoglu, A., Neyerlin, K.C., Mukundan, R., Ahluwalia, R.K., Cullen, D.A., More, K.L., Weber, A.Z., Myers, D.J., Recent developments in catalyst-related PEM fuel cell durability. *Curr. Opin. Electrochem.*, 21, 192–200, 2020. https://doi.org/10.1016/j.coelec.2020.02.007.
47. Lin, R., Che, L., Shen, D., Cai, X., High durability of Pt-Ni-Ir/C ternary catalyst of PEMFC by stepwise reduction synthesis. *Electrochim. Acta.*, 330, 135251, 2020. https://doi.org/10.1016/j.electacta.2019.135251.
48. Jiao, K., Xuan, J., Du, Q., Bao, Z., Xie, B., Wang, B., Zhao, Y., Fan, L., Wang, H., Hou, Z., Huo, S., Brandon, N.P., Yin, Y., Guiver, M.D., Designing the next generation of proton-exchange membrane fuel cells. *Nature*, 595, 361–369, 2021. https://doi.org/10.1038/s41586-021-03482-7.
49. Chen, M., Zhao, C., Sun, F., Fan, J., Li, H., Wang, H., Research progress of catalyst layer and interlayer interface structures in membrane electrode assembly (MEA) for proton exchange membrane fuel cell (PEMFC) system. *ETransportation*, 5, 100075, 2020. https://doi.org/10.1016/j.etran.2020.100075.
50. Xie, M., Chu, T., Wang, T., Wan, K., Yang, D., Li, B., Ming, P., Zhang, C., Preparation, performance and challenges of catalyst layer for proton exchange membrane fuel cell. *Membranes (Basel)*, 11, 1–33, 2021. https://doi.org/10.3390/membranes11110879.
51. Tian, X., Zhao, X., Su, Y.-Q., Wang, L., Wang, H., Dang, D., Chi, B., Liu, H., Hensen, E.J.M., Lou, (David) X.W., Xia, B.Y., Engineering bunched Pt-Ni alloy nanocages for efficient oxygen reduction in practical fuel cells. *Science*, 366, 850–856, 2019. https://doi.org/10.1126/science.aaw7493.
52. Li, M., Zhao, Z., Cheng, T., Fortunelli, A., Chen, C.-Y., Yu, R., Zhang, Q., Gu, L., Merinov, B.V., Lin, Z., Zhu, E., Yu, T., Jia, Q., Guo, J., Zhang, L., Goddard, W.A., Huang, Y., Duan, X., Ultrafine jagged platinum nanowires

enable ultrahigh mass activity for the oxygen reduction reaction. *Science*, 354, 1414–1419, 2016. https://doi.org/10.1126/science.aaf9050.
53. Liu, J., Li, Y., Wu, Z., Ruan, M., Song, P., Jiang, L., Wang, Y., Sun, G., Xu, W., Pt0.61Ni/C for high-efficiency cathode of fuel cells with superhigh platinum utilization. *J. Phys. Chem. C.*, 122, 14691–14697, 2018. https://doi.org/10.1021/acs.jpcc.8b03966.
54. Wu, Z.P., Caracciolo, D.T., Maswadeh, Y., Wen, J., Kong, Z., Shan, S., Vargas, J.A., Yan, S., Hopkins, E., Park, K., Sharma, A., Ren, Y., Petkov, V., Wang, L., Zhong, C.J., Alloying–realloying enabled high durability for Pt–Pd-3d-transition metal nanoparticle fuel cell catalysts. *Nat. Commun.*, 12, 1–14, 2021. https://doi.org/10.1038/s41467-021-21017-6.
55. Chen, C., Kang, Y., Huo, Z., Zhu, Z., Huang, W., Xin, H.L., Snyder, J.D., Li, D., Herron, J.A., Mavrikakis, M., Chi, M., More, K.L., Li, Y., Markovic, N.M., Somorjai, G.A., Yang, P., Stamenkovic, V.R., Highly crystalline multimetallic nanoframes with three-dimensional electrocatalytic surfaces. *Science*, 343, 1339–1343, 2014. https://doi.org/10.1126/science.1249061.
56. Prabal Sapkota, C.C., Boyer, C., Dutta, R., Aguey-Zinsou, K.-F., Planar polymer electrolyte membrane fuel cells: Powering portable devices from hydrogen. *Sustain. Energy Fuels*, 4, 439, 2020. https://doi.org/10.1039/c9se00861f.
57. Chandran, P. and Ramaprabhu, S., Catalytic performance of non-platinum-based hybrid carbon hetero-structure for oxygen reduction and hydrogen oxidation reactions in proton exchange membrane fuel cell. *Int. J. Hydrogen Energy*, 43, 18477–18487, 2018. https://doi.org/10.1016/j.ijhydene.2018.08.066.
58. Wang, Y., Yuan, H., Martinez, A., Hong, P., Xu, H., Bockmiller, F.R., Polymer electrolyte membrane fuel cell and hydrogen station networks for automobiles: Status, technology, and perspectives. *Adv. Appl. Energy*, 2, 100011, 2021. https://doi.org/10.1016/j.adapen.2021.100011.
59. Chen, Q., Niu, Z., Li, H., Jiao, K., Wang, Y., ScienceDirect recent progress of gas diffusion layer in proton exchange membrane fuel cell: Two-phase flow and material properties. *Int. J. Hydrogen Energy*, 46, 8640–8671, 2020. https://doi.org/10.1016/j.ijhydene.2020.12.076.
60. Xu, Z., Qiu, D., Yi, P., Peng, L., Lai, X., Towards mass applications: A review on the challenges and developments in metallic bipolar plates for PEMFC. *Prog. Nat. Sci. Mater. Int.*, 30, 815–824, 2020. https://doi.org/10.1016/j.pnsc.2020.10.015.
61. Hou, M., Shao, Z., Yi, B., Technological approaches to increasing specific power of vehicular fuel cell stacks. *Chinese J. Eng. Sci.*, 21, 84, 2019. https://doi.org/10.15302/j-sscae-2019.03.002.
62. Adloo, A., Sadeghi, M., Masoomi, M., Naja, H., High performance polymeric bipolar plate based on polypropylene/graphite/graphene/nano-carbon black composites for PEM fuel cells. *Renew. Energy*, 99, 867–874, 2016. https://doi.org/10.1016/j.renene.2016.07.062.

63. Naji, A., Krause, B., Ameli, A., State, W., Hybrid conductive filler/polycarbonate composites with enhanced electrical and thermal conductivities for bipolar plate applications. *Polym. Composites.*, 40, 3189–3198, 2019. https://doi.org/10.1002/pc.25169.
64. Wang, Y., Sun, Z.Y., Yang, L., Enhancement effects of the obstacle arrangement and gradient height distribution in serpentine flow-field on the performances of a PEMFC. *Energy Convers. Manag.*, 252, 115077, 2022. https://doi.org/10.1016/j.enconman.2021.115077.
65. Lee, E., Kim, J., Kim, T., Song, H., Kim, J., Park, S., Jeong, T., Yun, S., Lee, J., Goo, J., Hyuk, J., Gyu, B., Chun, H., Keun, P., Gil, C., Kim, Y., Enhanced corrosion resistance and fuel cell performance of Al1050 bipolar plate coated with TiN/Ti double layer. *Energy Convers. Manag.*, 75, 727–733, 2013. https://doi.org/10.1016/j.enconman.2013.08.041.
66. Fawwaz, N., Husaini, T., Bakar, A., Herianto, E., Ramli, W., Daud, W., Coating of stainless steel and titanium bipolar plates for anticorrosion in PEMFC: A review. *Int. J. Hydrogen Energy*, 42, 9135–9148, 2015. https://doi.org/10.1016/j.ijhydene.2016.06.241.
67. Zhao, Y., Wei, L., Yi, P., Peng, L., Influence of Cr-C film composition on electrical and corrosion properties of 316L stainless steel as bipolar plates for PEMFCs. *Int. J. Hydrogen Energy*, 41, 1142–1150, 2015. https://doi.org/10.1016/j.ijhydene.2015.10.047.
68. Forouzanmehr, M., Kashyzadeh, K.R., Borjali, A., Ivanov, A., Jafarnode, M., Gan, T.H., Wang, B., Chizari, M., Detection and analysis of corrosion and contact resistance faults of TiN and CrN coatings on 410 stainless steel as bipolar plates in PEM fuel cells. *Sensors*, 22, 2022. https://doi.org/10.3390/s22030750.
69. Bi, J., Yang, J., Liu, X., Wang, D., Yang, Z., Liu, G., Wang, X., Development and evaluation of nitride coated titanium bipolar plates for PEM fuel cells. *Int. J. Hydrogen Energy*, 46, 1144–1154, 2021. https://doi.org/10.1016/j.ijhydene.2020.09.217.
70. Li, H., Guo, P., Zhang, D., Liu, L., Wang, Z., Ma, G., Xin, Y., Ke, P., Saito, H., Wang, A., Interface-induced degradation of amorphous carbon films/stainless steel bipolar plates in proton exchange membrane fuel cells. *J. Power Sources*, 469, 228269, 2020. https://doi.org/10.1016/j.jpowsour.2020.228269.
71. Yang, L.X., Liu, R.J., Wang, Y., Liu, H.J., Zeng, C.L., Fu, C., Growth of nanocrystalline β-Nb2N coating on 430 ferritic stainless steel bipolar plates of PEMFCs by disproportionation reaction of Nb(IV) ions in molten salt. *Corros. Sci.*, 174, 108862, 2020. https://doi.org/10.1016/j.corsci.2020.108862.
72. Deyab, M.A. and Mele, G., Stainless steel bipolar plate coated with polyaniline/Zn-porphyrin composites coatings for proton exchange membrane fuel cell. *Sci. Rep.*, 10, 1–8, 2020. https://doi.org/10.1038/s41598-020-60288-9.
73. Vlaskin, M.S., Grigorenko, A.V., Shkolnikov, E.I., Ilyukhin, A.S., Gold-plated titanium vs carbon-implanted titanium as material for bipolar plates

in PEM fuel cells. *Surf. Rev. Lett.*, 26, 1–5, 2019. https://doi.org/10.1142/S0218625X19500380.
74. Lu, C., Shi, F., Jin, J., Peng, X., Study on the properties of vertical carbon nanotube films grown on stainless steel bipolar plates. *Materials (Basel).*, 16, 6, 2019, doi: 10.3390/ma12060899.
75. Yumiya, H., Kizaki, M., Asai, H., Toyota fuel cell system (TFCS). *World Electr. Veh. J.*, 7, 1, 85–92, 2015, doi: 10.3390/wevj7010085.
76. Dubau, L., Castanheira, L., Maillard, F., Lottin, O., Maranzana, G., Dillet, J., Perrin, J., Moukheiber, E., De Moor, G., Bas, C., Flandin, L., Caqué, N., A review of PEM fuel cell durability: Materials degradation, local heterogeneities of aging and possible mitigation strategies. *WIRES Energy and Environment.*, 3, 540–560, 2014.
77. Zhang, Y., Liu, C., Wan, Z., Yang, C., Li, S., Tu, Z., Wu, M., Chen, Y., Zhou, W., Performance enhancement of PEM fuel cells with an additional outlet in the parallel flow field. *Processes*, 9, 2021. https://doi.org/10.3390/pr9112061.
78. Atyabi, S.A., Afshari, E., Three-dimensional multiphase model of proton exchange membrane fuel cell with honeycomb flow field at the cathode side. *J. Clean. Prod.*, 214, 738–748, 2019. https://doi.org/10.1016/j.jclepro.2018.12.293.
79. Cho, J.I.S., Neville, T.P., Trogadas, P., Meyer, Q., Wu, Y., Ziesche, R., Boillat, P., Cochet, M., Manzi-Orezzoli, V., Shearing, P., Brett, D.J.L., Coppens, M.O., Visualization of liquid water in a lung-inspired flow-field based polymer electrolyte membrane fuel cell via neutron radiography. *Energy*, 170, 14–21, 2019. https://doi.org/10.1016/j.energy.2018.12.143.
80. Guo, N., Leu, M.C., Koylu, U.O., Bio-inspired flow field designs for polymer electrolyte membrane fuel cells. *Int. J. Hydrog. Energy*, 39, 36, 21185–21195, 2014, https://doi.org/10.1016/j.ijhydene.2014.10.069.
81. Huang P., *et al.*, Stainless steel bipolar plate fuel cell with different flow field structures prepared by laser additive manufacturing. *Int. J. Heat Mass Transf.*, 183, 122186, 2022, doi: 10.1016/j.ijheatmasstransfer.2021.122186
82. Kodama, K., Nagai, T., Kuwaki, A., Jinnouchi, R., Morimoto, Y., Challenges in applying highly active Pt-based nanostructured catalysts for oxygen reduction reactions to fuel cell vehicles. *Nat. Nanotechnol.*, 16, 140–147, 2021. https://doi.org/10.1038/s41565-020-00824-w.
83. Lim, S.Y., Martin, S., Gao, G., Dou, Y., Simonsen, S.B., Jensen, J.O., Li, Q., Norrman, K., Jing, S., Zhang, W., Self-standing nanofiber electrodes with Pt–Co derived from electrospun zeolitic imidazolate framework for high temperature PEM fuel cells. *Adv. Funct. Mater.*, 31, 1–9, 2021. https://doi.org/10.1002/adfm.202006771.
84. Yartys, V.A., Lototskyy, M.V., Linkov, V., Pasupathi, S., Davids, M.W., Tolj, I., Radica, G., Denys, R.V., Eriksen, J., Taube, K., Bellosta von Colbe, J., Capurso, G., Dornheim, M., Smith, F., Mathebula, D., Swanepoel, D., Suwarno, S., Hydride4Mobility: An EU Horizon 2020 project on hydrogen powered fuel cell utility vehicles using metal hydrides in hydrogen storage and

refueling systems. *Int. J. Hydrogen Energy*, 46, 35896–35909, 2021. https://doi.org/10.1016/j.ijhydene.2021.01.190.
85. Sankir, N.D. and Sankir, M., Eds., *Hydrogen Storage Technologies*. Wiley, Scrivener Publishing, Salem, 2018.
86. Alaswad, A., Baroutaji, A., Achour, H., Carton, J., Al Makky, A., Olabi, A.G., Developments in fuel cell technologies in the transport sector. *Int. J. Hydrogen Energy*, 41, 16499–16508, 2016. https://doi.org/10.1016/j.ijhydene.2016.03.164.
87. Rivard, E., Trudeau, M., Zaghib, K., Hydrogen storage for mobility: A review. *Materials (Basel)*, 12, 2019. https://doi.org/10.3390/ma12121973.
88. Yahashi, H., Yamashita, A., Shigemitsu, N., Goto, S., Kida, K., Inou, T., Development of high-pressure hydrogen storage system for new FCV, in *SAE Technical Papers*, no. 2021, 2021. doi: 10.4271/2021-01-0741.
89. 2021 Mirai Product Information, *Toyota*, 2021. https://s3.amazonaws.com/toyota-cms-media/wp-content/uploads/2019/10/2021-Toyota-Mirai_Product-Info-Guide.pdf (accessed Mar. 14, 2022).
90. Aziz, M., Liquid hydrogen: A review on liquefaction, storage, transportation, and safety. *Energies*, 14, 18, 5917, 2021. doi: 10.3390/en14185917.
91. Moreno-Blanco, J., Petitpas, G., Espinosa-loza, F., Elizalde-Blancas, F., Martinez-Frias, J., Aceves, S.M., The fill density of automotive cryo-compressed hydrogen vessels. *Int. J. Hydrogen Energy*, 44, 1010–1020, 2018. https://doi.org/10.1016/j.ijhydene.2018.10.227.
92. Brunner, T., Cryo-compressed hydrogen storage. *BMW Hydrog. Storage Work.*, Washington D.C., 2011.
93. Miller, E.L., Thompson, S.T., Randolph, K., Hulvey, Z., Rustagi, N., Satyapal, S., US Department of Energy hydrogen and fuel cell technologies perspectives, in: *MRS Bulletin*, pp. 57–64, 2020. https://doi.org/10.1557/mrs.2019.312.
94. Office of Energy Efficiency & Renewable Energy. Materials-based hydrogen storage. *US DOE*, 2020. https://www.energy.gov/eere/fuelcells/materials-based-hydrogen-storage (accessed Mar. 14, 2022).
95. Dillon, A.C., Jones, K.M., Bekkedahl, T.A., Kiang, C.H., Bethune, D.S., Heben, M.J., Storage of hydrogen in single-walled carbon nanotubes. *Nature*, 386, 377–379, 1997. https://doi.org/10.1038/386377a0.
96. Lyu, J., Kudiiarov, V., Lider, A., An overview of the recent progress in modifications of carbon nanotubes for hydrogen adsorption. *Nanomaterials*, 10, 255, 2020. https://doi.org/10.3390/nano10020255.
97. Rowsell, J.L.C., Millward, A.R., Park, K.S., Yaghi, O.M., Hydrogen sorption in functionalized metal-organic frameworks. *J. Am. Chem. Soc.*, 126, 5666–5667, 2004. https://doi.org/10.1021/ja049408c.
98. Hydrogen and Fuel Cell Technologies Office. *Materials-Based Hydrogen Storage*, 2020.
99. Dafedar, A.A., Verma, S.S., Yadav, A., Hydrogen storage techniques for stationary and mobile applications: A review, in *Recent Advances in Sustainable*

Technologies, K. Jha, P. Gulati, and U. K. Tripathi (Eds.), pp. 29–39, Springer, Singapore, 2021. doi: https://doi.org/10.1007/978-981-16-0976-3.
100. Millange, F., Serre, C., Guillou, N., Férey, G., Walton, R.I., Structural effects of solvents on the breathing of metal–organic frameworks: An *in situ* diffraction study. *Angew. Chemie Int. Ed.*, 47, 4100–4105, 2008. https://doi.org/10.1002/anie.200705607.
101. He, Y., Chen, F., Li, B., Qian, G., Zhou, W., Chen, B., Porous metal–organic frameworks for fuel storage. *Coord. Chem. Rev.*, 373, 167–198, 2018. https://doi.org/10.1016/j.ccr.2017.10.002.
102. Witman, M., Ling, S., Gladysiak, A., Stylianou, K.C., Smit, B., Slater, B., Haranczyk, M., Rational design of a low-cost, high-performance metal–organic framework for hydrogen storage and carbon capture. *J. Phys. Chem. C.*, 121, 1171–1181, 2017. https://doi.org/10.1021/acs.jpcc.6b10363.
103. Ahmed, A., Seth, S., Purewal, J., Wong-Foy, A.G., Veenstra, M., Matzger, A.J., Siegel, D.J., Exceptional hydrogen storage achieved by screening nearly half a million metal-organic frameworks. *Nat. Commun.*, 10, 1568, 2019. https://doi.org/10.1038/s41467-019-09365-w.
104. Teichmann, D., Arlt, W., Wasserscheid, P., Liquid organic hydrogen carriers as an efficient vector for the transport and storage of renewable energy. *Int. J. Hydrogen Energy*, 37, 18118–18132, 2012. https://doi.org/10.1016/j.ijhydene.2012.08.066.
105. Verevkin, S.P., Andreeva, I.V., Konnova, M.E., Portnova, S.V., Zherikova, K.V., Pimerzin, A.A., Paving the way to the sustainable hydrogen storage: Thermochemistry of amino-alcohols as precursors for liquid organic hydrogen carriers. *J. Chem. Thermodyn.*, 163, 106610, 2021. https://doi.org/10.1016/j.jct.2021.106610.
106. Singh, R., Singh, M., Gautam, S., Hydrogen economy, energy, and liquid organic carriers for its mobility. *Mater. Today Proc.*, 46, 5420–5427, 2021. https://doi.org/10.1016/j.matpr.2020.09.065.
107. Ren, J., Musyoka, N.M., Langmi, H.W., Mathe, M., Liao, S., Current research trends and perspectives on materials-based hydrogen storage solutions: A critical review. *Int. J. Hydrogen Energy*, 42, 289–311, 2017. https://doi.org/10.1016/j.ijhydene.2016.11.195.
108. Andersson, J. and Grönkvist, S., Large-scale storage of hydrogen. *Int. J. Hydrogen Energy*, 44, 11901–11919, 2019. https://doi.org/10.1016/j.ijhydene.2019.03.063.
109. Bellosta von Colbe, J., Ares, J.-R., Barale, J., Baricco, M., Buckley, C., Capurso, G., Gallandat, N., D.M. Grant, D.M., Guzik, M.N., Jacob, I., Jensen, E.H., Jensen, T., Jepsen, J., Klassen, T., Lototskyy, M.V., Manickam, K., Montone, A., Puszkiel, J., Sartori, S., Sheppard, D.A., Stuart, A., Walker, G., Webb, C.J., Yang, H., Yartys, V., Züttel, A., Dornheim, M., Application of hydrides in hydrogen storage and compression: Achievements, outlook and perspectives. *Int. J. Hydrogen Energy*, 44, 7780–7808, 2019. https://doi.org/10.1016/j.ijhydene.2019.01.104.

110. Shi, R., Yan, H., Zhang, J., Gao, H., Zhu, Y., Liu, Y., Hu, X., Zhang, Y., Li, L., Vacancy-mediated hydrogen spillover improving hydrogen storage properties and air stability of metal hydrides. *Small*, 17, 2100852, 2021. https://doi.org/10.1002/smll.202100852.
111. Schneemann, A., White, J.L., Kang, S., Jeong, S., Wan, L.F., Cho, E.S., Heo, T.W., Prendergast, D., Urban, J.J., Wood, B.C., Allendorf, M.D., Stavila, V., Nanostructured metal hydrides for hydrogen storage. *Chem. Rev.*, 118, 10775–10839, 2018. https://doi.org/10.1021/acs.chemrev.8b00313.
112. Sakintuna, B., Lamari-Darkrim, F., Hirscher, M., Metal hydride materials for solid hydrogen storage: A review. *Int. J. Hydrogen Energy*, 32, 1121–1140, 2007. https://doi.org/10.1016/j.ijhydene.2006.11.022.
113. Lototskyy, M., Tolj, I., Klochko, Y., Davids, M.W., Swanepoel, D., Linkov, V., Metal hydride hydrogen storage tank for fuel cell utility vehicles. *Int. J. Hydrogen Energy*, 45, 7958–7967, 2020. https://doi.org/10.1016/j.ijhydene.2019.04.124.
114. Grochala, W. and Edwards, P.P., Thermal decomposition of the non-interstitial hydrides for the storage and production of hydrogen. *Chem. Rev.*, 104, 1283–1316, 2004. https://doi.org/10.1021/cr030691s.
115. Imamura, H., Masanari, K., Kusuhara, M., Katsumoto, H., Sumi, T., Sakata, Y., High hydrogen storage capacity of nanosized magnesium synthesized by high energy ball-milling. *J. Alloys Compd.*, 386, 211–216, 2005. https://doi.org/10.1016/j.jallcom.2004.04.145.
116. Ren, C., Fang, Z.Z., Zhou, C., Lu, J., Ren, Y., Zhang, X., Hydrogen storage properties of magnesium hydride with V-based additives. *J. Phys. Chem. C.*, 118, 21778–21784, 2014. https://doi.org/10.1021/jp504766b.
117. Møller, K.T., Sheppard, D., Ravnsbæk, D.B., Buckley, C.E., Akiba, E., Li, H.W., Jensen, T.R., Complex metal hydrides for hydrogen, thermal and electrochemical energy storage. *Energies*, 10, 2017. https://doi.org/10.3390/en10101645.
118. Sankir, N.D. and Sankir, M., Eds., *Hydrogen Production Technologies*. Wiley, Scrivener Publishing, Salem, 2017.
119. Sankir, M., Semiz, L., Serin, R.B., Sankir, N.D., Baker, D., Hydrogen generation from chemical hydrides. *Adv. Catal. Mater.*, 143–192, 2015. https://doi.org/10.1002/9781118998939.ch5.
120. Sankir, M., Serin, R.B., Semiz, L., Sankir, N.D., Unusual behavior of dynamic hydrogen generation from sodium borohydride. *Int. J. Hydrogen Energy*, 39, 2608–2613, 2014. https://doi.org/10.1016/j.ijhydene.2013.12.011.
121. Colak, T.O. *et al.*, Efficient hydrogen generation with Co3O4@TiO2-g-C3N4 composite catalyst via catalytic NaBH4 hydrolysis. *Catal. Letters*, 152, 9, 2779–2788, 2022, doi: 10.1007/s10562-021-03848-6.
122. Shabunya, S.I., Minkina, V.G., Kalinin, V.I., Sanki, N.D., Altaf, C.T., Russian Text © The Author(s), 2021. *Kinet. Catal.*, 62, 305–315, 2021. https://doi.org/10.1134/S0023158421030083.

123. Semiz, L., Abdullayeva, N., Sankir, M., Nanoporous Pt and Ru catalysts by chemical dealloying of Pt-Al and Ru-Al alloys for ultrafast hydrogen generation. *J. Alloys Compd.*, 744, 110–115, 2018. https://doi.org/10.1016/j.jallcom.2018.02.082.
124. Lee, H., Choi, W.I., Ihm, J., Combinatorial search for optimal hydrogen-storage nanomaterials based on polymers. *Phys. Rev. Lett.*, 97, 56104, 2006. https://doi.org/10.1103/PhysRevLett.97.056104.
125. Skipper, C.V.J., Hamaed, A., Antonelli, D.M., Kaltsoyannis, N., The Kubas interaction in M(ii) (M = Ti{,} V{,} Cr) hydrazine-based hydrogen storage materials: A DFT study. *Dalt. Trans.*, 41, 8515–8523, 2012. https://doi.org/10.1039/C2DT30383C.
126. Chung, C., Ihm, J., Lee, H., Recent progress on Kubas-type hydrogen-storage nanomaterials: From theories to experiments. *J. Korean Phys. Soc.*, 66, 1649–1655, 2015. https://doi.org/10.3938/jkps.66.1649.
127. Boateng, E. and Chen, A., Recent advances in nanomaterial-based solid-state hydrogen storage. *Mater. Today Adv.*, 6, 100022, 2020. https://doi.org/10.1016/j.mtadv.2019.100022.
128. D. P. and S. Ramaprabhu, Hydrogen storage in platinum decorated hydrogen exfoliated graphene sheets by spillover mechanism. *Phys. Chem. Chem. Phys.*, 16, 26725–26729, 2014. https://doi.org/10.1039/C4CP04214J.
129. Morris, L., Trudeau, M.L., Reed, D., Book, D., Antonelli, D.M., Thermodynamically neutral Kubas-type hydrogen storage using amorphous Cr(iii) alkyl hydride gels. *Phys. Chem. Chem. Phys.*, 17, 9480–9487, 2015. https://doi.org/10.1039/C5CP00412H.
130. Akay, T.E., Abdullayeva, N., Sankir, M., Demirci Sankir, N., On-board hydrogen powered proton exchange membrane fuel cells. *ECS Trans.*, 75, 14, 511–513, 2016, doi: 10.1149/07514.0511ecst.
131. Carpenter, I., Edwards, N., Ellis, S., Frost, J., Golunski, S., Van Keulen, N., Petch, M., Pignon, J., Reinkingh, J., On-board hydrogen generation for PEM fuel cells in automotive applications. *SAE Tech. Pap.* (1999). https://doi.org/10.4271/1999-01-1320.
132. Avcı, A.K., Önsan, Z.İ., Trimm, D.L., On-board hydrogen generation for fuel cell-powered vehicles: The use of methanol and propane. *Top. Catal.* 22, 359–367, 2003. https://doi.org/10.1023/A:1023504826480.
133. Elitzur, S., Rosenband, V., Gany, A., On-board hydrogen production for auxiliary power in passenger aircraft. *Int. J. Hydrogen Energy*, 42, 14003–14009, 2017. https://doi.org/10.1016/j.ijhydene.2017.02.037.
134. Corsi, J.S., Fu, J., Wang, Z., Lee, T., Ng, A.K., Detsi, E., Hierarchical bulk nanoporous aluminum for on-site generation of hydrogen by hydrolysis in pure water and combustion of solid fuels. *ACS Sustain. Chem. Eng.*, 7, 11194–11204, 2019. https://doi.org/10.1021/acssuschemeng.9b00481.
135. Kim, T. and Kwon, S., Design and development of a fuel cell-powered small unmanned aircraft. *Int. J. Hydrogen Energy*, 37, 615–622, 2012. https://doi.org/10.1016/j.ijhydene.2011.09.051.

136. N.D. Sankir, M. Sankir, L. Semiz, and R.B. Serin, A cartridge structure designed for generation of hydrogen gas, US10272405B2, 2015 [Online]. Available: https://patents.google.com/patent/US10272405B2/en
137. IRENA. *Green Hydrogen: A Guide to Policy Making*, p. 8, International Renewable Energy Agency, Abu Dhabi, 2020.
138. Howarth, R.W. and Jacobson, M.Z., How green is blue hydrogen? *Energy Sci. Eng.*, 9, 1676–1687, 2021. https://doi.org/10.1002/ese3.956.
139. Gong, J., A commentary of green hydrogen in MIT technology review 2021. *Fundam. Res.*, 1, 848–850, 2021. https://doi.org/10.1016/j.fmre.2021.11.013.
140. Wang, M., Wang, G., Sun, Z., Zhang, Y., Xu, D., Review of renewable energy-based hydrogen production processes for sustainable energy innovation. *Glob. Energy Interconnect.*, 2, 436–443, 2019. https://doi.org/10.1016/j.gloei.2019.11.019.
141. Osman, A.I., Mehta, N., Elgarahy, A.M., Hefny, M., Al-Hinai, A., Al-Muhtaseb, A.H., Rooney, D.W., Hydrogen production, storage, utilisation and environmental impacts: A review. Springer International Publishing, 2021. https://doi.org/10.1007/s10311-021-01322-8.
142. Altaf, C.T., Faraji, M., Kumtepe, A., Abdullayeva, N., Yilmaz, N., Karagoz, E., Kurt, H., Sankir, M., Sankir, N.D., Highly efficient 3D-ZnO nanosheet photoelectrodes for solar-driven water splitting: Chalcogenide nanoparticle sensitization and mathematical modeling. *J. Alloys Compd.*, 154472, 2020. https://doi.org/10.1016/j.jallcom.2020.154472.
143. Altaf, C.T., Yolacan, D., Sankir, N.D., Decoration of 3D ZnO nanoelectrodes with CuInS2 for solar water splitting. *Mater. Lett.*, 236, 710–714, 2019. https://doi.org/10.1016/j.matlet.2018.11.056.
144. Tuc Altaf, C., Sahsuvar, N.S., Abdullayeva, N., Coskun, O., Kumtepe, A., Karagoz, E., Sankir, M., Demirci Sankir, N., Inverted configuration of Cu(In,Ga)S2/In2S3on 3D-ZnO/ZnSnO3 bilayer system for highly efficient photoelectrochemical water splitting. *ACS Sustain. Chem. Eng.*, 8, 15209–15222, 2020. https://doi.org/10.1021/acssuschemeng.0c04846.
145. Altaf, C.T., Yolacan, D., Sankir, N.D., Decoration of 3D ZnO nanoelectrodes with CuInS2 for solar water splitting. *Mater. Lett.*, 236, 710–714, 2019. https://doi.org/10.1016/j.matlet.2018.11.056.
146. Altaf, C.T., Abdullayeva, N., Coskun, O., Kumtepe, A., Yildirim, I.D., Erdem, E., Liu, M., Bozbey, A., Agar, E., Sankir, M., Sankir, N.D., Efficiency enhancement in photoelectrochemical water splitting: Defect passivation and boosted charge transfer kinetics of zinc oxide nanostructures via chalcopyrite/chalcogenide mix sensitization. *Phys. Rev. Mater.*, 5, 125403, 1–14, 2021. https://doi.org/10.1103/PhysRevMaterials.5.125403.
147. Abdullayeva, N., Altaf, C.T., Mintas, M., Ozer, A., Sankir, M., Kurt, H., Sankir, N.D., Investigation of strain effects on photoelectrochemical performance of flexible ZnO electrodes. *Sci. Rep.*, 9, 1–14, 2019. https://doi.org/10.1038/s41598-019-47546-1.

148. Troppmann, S. and König, B., Functionalized membranes for photocatalytic hydrogen production. *Chem. - A Eur. J.*, 20, 14570–14574, 2014. https://doi.org/10.1002/chem.201404480.
149. Tuc Altaf, Ç. and Demirci Sankir, N., Colloidal synthesis of CuInS2 nanoparticles: Crystal phase design and thin film fabrication for photoelectrochemical solar cells. *Int. J. Hydrogen Energy*, 44, 18712–18723, 2019. https://doi.org/10.1016/j.ijhydene.2018.11.047.
150. Landman, A., Dotan, H., Shter, G.E., Wullenkord, M., Houaijia, A., Maljusch, A., Grader, G.S., Rothschild, A., Photoelectrochemical water splitting in separate oxygen and hydrogen cells. *Nat. Mater.*, 16, 646–651, 2017. https://doi.org/10.1038/nmat4876.
151. Komurcu, P., Can, E.K., Aydin, E., Semiz, L., Sankir, M., Materials science in semiconductor processing cost-effective fabrication of nanostructured zinc oxide-based electrodes for photoelectrochemical water splitting. *Mater. Sci. Semicond. Process.*, 42, 159–164, 2016. https://doi.org/10.1016/j.mssp.2015.07.054.
152. Altaf, C.T., Abdullayeva, N., Sankir, N.D., Sankir, N.D., Copper-based chalcopyrite and kesterite materials for solar hydrogen generation, in *Photoelectrochemical Solar Cells*, 1st ed., N.D. Sankir and M. Sankir (Eds.), pp. 251–303, Wiley, Scrivener Publishing, Salem, 2018. doi: https://doi.org/10.1002/9781119460008.ch7.
153. Frowijn, L.S.F. and van Sark, W.G.J.H.M., Analysis of photon-driven solar-to-hydrogen production methods in the Netherlands. *Sustain. Energy Technol. Assessments*, 48, 101631, 2021. https://doi.org/10.1016/j.seta.2021.101631.
154. Nishiyama, H., Yamada, T., Nakabayashi, M., Maehara, Y., Yamaguchi, M., Kuromiya, Y., Nagatsuma, Y., Tokudome, H., Akiyama, S., Watanabe, T., Narushima, R., Okunaka, S., Shibata, N., Takata, T., Hisatomi, T., Domen, K., Photocatalytic solar hydrogen production from water on a 100-m^2 scale. *Nature*, 598, 304–307, 2021. https://doi.org/10.1038/s41586-021-03907-3.
155. Sandhana, L., New technique for turning sunshine and water into hydrogen fuel. *New Atlas*, 2013. https://newatlas.com/sunshine-and-water-into-fuel/28563/ (accessed Mar. 14, 2022).
156. Sankir, M., Semiz, L., Serin, R.B., ScienceDirect hydrogen generation from nanoflower platinum films. *Int. J. Hydrogen Energy*, 40, 8522–8529, 2015. https://doi.org/10.1016/j.ijhydene.2015.04.137.
157. Serin, R.B., Abdullayeva, N., Sankir, M., Dealloyed ruthenium film catalysts for hydrogen generation from chemical hydrides. *Materials (Basel)*, 10, 1–10, 2017. https://doi.org/10.3390/ma10070738.
158. Tembhurne, S., Nandjou, F., Haussener, S., A thermally synergistic photo-electrochemical hydrogen generator operating under concentrated solar irradiation. *Nat. Energy*, 4, 5, 399–407, 2019, doi: 10.1038/s41560-019-0373-7.
159. Oldenbroek, V., Wijtzes, S., Blok, K., van Wijk, A.J.M., Fuel cell electric vehicles and hydrogen balancing 100 percent renewable and integrated national

transportation and energy systems. *Energy Convers. Manag. X.*, 9, 100077, 2021. https://doi.org/10.1016/j.ecmx.2021.100077.

160. Das, H.S., Tan, C.W., Yatim, A.H.M., Fuel cell hybrid electric vehicles: A review on power conditioning units and topologies. *Renew. Sustain. Energy Rev.*, 76, 268–291, 2017. https://doi.org/10.1016/j.rser.2017.03.056.
161. Mekhilef, S., Saidur, R., Safari, A., Comparative study of different fuel cell technologies. *Renew. Sustain. Energy Rev.*, 16, 981–989, 2012. https://doi.org/10.1016/j.rser.2011.09.020.
162. Wilberforce, T., El-Hassan, Z., Khatib, F.N., Al Makky, A., Baroutaji, A., Carton, J.G., Olabi, A.G., Developments of electric cars and fuel cell hydrogen electric cars. *Int. J. Hydrogen Energy*, 42, 25695–25734, 2017. https://doi.org/10.1016/j.ijhydene.2017.07.054.
163. Uzunoglu, M. and Alam, M.S., Dynamic modeling, design and simulation of a PEM fuel cell/ultra-capacitor hybrid system for vehicular applications. *Energy Convers. Manag.*, 48, 1544–1553, 2007. https://doi.org/10.1016/j.enconman.2006.11.014.
164. Islam, M.R., Shabani, B., Rosengarten, G., Andrews, J., The potential of using nanofluids in PEM fuel cell cooling systems: A review. *Renew. Sustain. Energy Rev.*, 48, 523–539, 2015. https://doi.org/10.1016/j.rser.2015.04.018.
165. Inal, O.B. and Deniz, C., Assessment of fuel cell types for ships: Based on multi-criteria decision analysis. *J. Clean. Prod.*, 265, 121734, 2020. https://doi.org/10.1016/j.jclepro.2020.121734.
166. Shilaja, C. *et al.*, Design and analysis of global optimization methods for proton exchange membrane fuel cell powered electric vehicle system with single switch DC-DC converter. *Mater. Today Proc.*, 52, xxxx, 2057–2064, 2022, doi: 10.1016/j.matpr.2021.12.204.
167. Authayanun, S., Im-Orb, K., Arpornwichanop, A., A review of the development of high temperature proton exchange membrane fuel cells. *Cuihua Xuebao/Chinese J. Catal.*, 36, 473–483, 2015. https://doi.org/10.1016/S1872-2067(14)60272-2.
168. Cacciola, G., Antonucci, V., Freni, S., Technology up date and new strategies on fuel cells. *J. Power Sources*, 100, 67–79, 2001. https://doi.org/10.1016/S0378-7753(01)00884-9.
169. Ogungbemi, E., Wilberforce, T., Ijaodola, O., Thompson, J., Olabi, A.G., Selection of proton exchange membrane fuel cell for transportation. *Int. J. Hydrogen Energy*, 46, 30625–30640, 2021. https://doi.org/10.1016/j.ijhydene.2020.06.147.
170. Cheng, S., Xu, L., Li, J., Fang, C., Hu, J., Ouyang, M., Development of a PEM fuel cell city bus with a hierarchical control system. *Energies*, 9, 6, 2016, doi: 10.3390/en9060417.
171. Xu, L., Ouyang, M., Li, J., Yang, F., Lu, L., Hua, J., Optimal sizing of plug-in fuel cell electric vehicles using models of vehicle performance and system cost. *Appl. Energy.*, 103, 477–487, 2013. https://doi.org/10.1016/j.apenergy.2012.10.010.

172. Radica, G., Tolj, I., Markota, D., Lototskyy, M.V., Pasupathi, S., Yartys, V., Control strategy of a fuel-cell power module for electric forklift. *Int. J. Hydrogen Energy*, 46, 35938–35948, 2021. https://doi.org/10.1016/j.ijhydene.2021.01.225.
173. Koebler, J., Hyundai becomes first company to mass produce hydrogen fuel cell cars, *USNews.com*, 2015. https://www.usnews.com/news/articles/2013/02/26/hyundai-becomes-first-company-to-mass-produce-hydrogen-fuel-cell-cars (accessed Mar. 14, 2022).
174. Kane, M., Hydrogen fuel cell car sales in 2019 improved to 7,500 globally. *Insideevs.com*, 2020. https://insideevs.com/news/397240/hydrogen-fuel-cell-sales-2019-7500-globally/ (accessed Mar. 14, 2022).
175. Xu, J., Zhang, C., Wan, Z., Chen, X., Chan, S.H., Tu, Z., Progress and perspectives of integrated thermal management systems in PEM fuel cell vehicles: A review. *Renew. Sustain. Energy Rev.*, 155, 111908, 2022. https://doi.org/10.1016/j.rser.2021.111908.
176. Matsunaga, M., Fukushima, T., Ojima, K., Powertrain system of Honda FCX clarity fuel cell vehicle, *24th Int. Batter. Hybrid Fuel Cell Electr. Veh. Symp. Exhib. 2009*, EVS 24.4, pp. 2714–2723, 2009.
177. Hong, B.K. and Kim, S.H., (Invited) recent advances in fuel cell electric vehicle technologies of Hyundai. *ECS Trans.*, 86, 3–11, 2018. https://doi.org/10.1149/08613.0003ecst.
178. Chen, Q., Zhang, G., Zhang, X., Sun, C., Jiao, K., Wang, Y., Thermal management of polymer electrolyte membrane fuel cells: A review of cooling methods, material properties, and durability. *Appl. Energy.*, 286, 116496, 2021. https://doi.org/10.1016/j.apenergy.2021.116496.
179. Sando, Y., Research and development of fuel cell vehicles at Honda. *ECS Trans.*, 25, 211–224, 2019. https://doi.org/10.1149/1.3210573.
180. Wang, Y., Chen, K.S., Mishler, J., Cho, S.C., Adroher, X.C., A review of polymer electrolyte membrane fuel cells: Technology, applications, and needs on fundamental research. *Appl. Energy*, 88, 981–1007, 2011. https://doi.org/10.1016/j.apenergy.2010.09.030.
181. Yoshida, T. and Kojima, K., Toyota MIRAI fuel cell vehicle and progress toward a future hydrogen society. *Electrochem. Soc. Interface*, 24, 45–49, 2015. https://doi.org/10.1149/2.F03152if.
182. Usai, L., Hung, C.R., Vásquez, F., Windsheimer, M., Burheim, O.S., Strømman, A.H., Life cycle assessment of fuel cell systems for light duty vehicles, current state-of-the art and future impacts. *J. Cleaner Prod.*, 280, 125086, 2021, doi: 10.1016/j.jclepro.2020.125086.
183. Konno, N., Mizuno, S., Nakaji, H., Ishikawa, Y., Development of compact and high-performance fuel cell stack. *SAE Int. J. Altern. Powertrains*, 4, 123–129, 2015. https://doi.org/10.4271/2015-01-1175.
184. Pollet, B.G., Kocha, S.S., Staffell, I., Current status of automotive fuel cells for sustainable transport. *Curr. Opin. Electrochem.*, 16, 90–95, 2019. https://doi.org/10.1016/j.coelec.2019.04.021.

185. "Toyota Mirai," *Toyota UK*, 2022. https://www.toyota.co.uk/hydrogen (accessed Mar. 14, 2022).
186. Wang, Y., Ruiz Diaz, D.F., Chen, K.S., Wang, Z., Adroher, X.C., Materials, technological status, and fundamentals of PEM fuel cells – A review. *Mater. Today*, 32, 178–203, 2020. https://doi.org/10.1016/j.mattod.2019.06.005.
187. Peng, H. *et al.*, A comparison of various universally applicable power distribution strategies for fuel cell hybrid trains utilizing component modeling at different levels of detail: From simulation to test bench measurement. *eTransportation*, 9, 2021, doi: 10.1016/j.etran.2021.100120.
188. Miller, S., World's first hydrogen train Coradia iLint honoured. *Alstom.com*, 2021. https://www.alstom.com/press-releases-news/2021/1/worlds-first-hydrogen-train-coradia-ilint-honoured (accessed Mar. 14, 2022).
189. Editorial. Siemens Mireo Plus H hydrogen train to be tested in Bavaria and Baden-Württemberg. *Urban Transport Magazine*, 2021. https://www.urban-transport-magazine.com/en/siemens-mireo-plus-h-hydrogen-train-to-be-tested-in-bavaria-and-baden-wurttemberg/ (accessed Mar. 14, 2022).
190. Fuel cell-powered hybrid multiple unit – Mireo. *Now-gmbh.de*, 2021. https://www.now-gmbh.de/en/projectfinder/x-emu/ (accessed Mar. 14, 2022).

7

Power Density and Durability in Fuel Cell Vehicles

H. Heidary[1] and M. Moein-Jahromi[2*]

[1]*Department of Energy Engineering, Qom University of Technology, Qom, Iran*
[2]*Department of Mechanical Engineering, Jahrom University, Jahrom, Iran*

Abstract

Proton exchange membrane fuel cell (PEMFC) systems have great potential to substitute with internal combustion engines (ICEs) in vehicles. The significant advantages are high efficiency, high driving range, low refueling time, and zero tank-to-wheel emission. Nonetheless, some unresolved concerns still exist, including the low volumetric power density, low durability under automotive conditions, and competitive cost. It is tried to investigate the two first issues in this chapter in two separate parts. The ultimate target for volumetric power density is about 9 kW L^{-1} for 2040, which is determined by Japan and the European Union, while the present status for one of the pioneer fuel cell vehicles (FCVs), Toyota Mirai-2020, is 5.5 kW L^{-1}. The ultimate target will be met by reducing the volume of the fuel cell (means reducing the cell thickness) on the one hand and increasing the generated power on the other hand. The first part of this chapter addresses the potent approaches for increasing volumetric power density through different kinds of flow field development. The leading solutions for cutting the volume are using the novel electrode with reduced thickness and decreasing the channel depth. Some of the key remedies for enhancing power are channel indentation, using a novel 3D mesh flow field to manage the liquid water without clogging the routes, and using metal foam instead of traditional flow fields. The second part of this chapter presents the degradation mechanisms imposed on PEMFC under automotive conditions. An automotive fuel cell must be capable of working under operation conditions such as idling, near to open-circuit voltage (OCV), cyclic load between two voltages, high power (high current), and finally cyclic start/stop. These automotive conditions degrade the fuel cell components, especially the membrane and catalyst layers, and lessen

**Corresponding author*: mmoein@jahromu.ac.ir

the fuel cell lifespan. Therefore, the second part of this chapter introduces the membrane and catalyst layers degradations through free radicals' attack, reducing electrochemical surface area (ECSA), platinum particle aggregations via Ostwald ripening, platinum dissolution inside the ionomer of catalyst layer and membrane, carbon corrosion of the platinum supports. Finally, some solutions to mitigate the degradation rate under automotive conditions, such as using new hybrid catalysts and energy management strategies to alleviate the cyclic load and start/stop, are also presented.

***Keywords*:** Volumetric power density, power enhancement, novel flow fields, automotive fuel cell degradation, degradation mechanisms, energy management strategies

7.1 Fuel Cell Performance and Power Density

7.1.1 Introduction

Proton exchange membrane fuel cell (PEMFC) systems have great potential to substitute with internal combustion engines (ICEs) in vehicles. The significant advantages are high efficiency, high driving range, low refueling time, and zero tank-to-wheel emission. Nonetheless, some unresolved concerns still exist, including the low volumetric power density, low durability under automotive conditions, and competitive cost.

The ultimate target for volumetric power density is about 9 kW L^{-1} for 2040, which is determined by Japan and the European Union, while the present status for one of the pioneer fuel cell vehicles (FCVs), Toyota Mirai-2020, is 5.5 kW L^{-1} [1]. The ultimate target will be met by reducing the volume of the fuel cell (means reducing the cell thickness) on the one hand and increasing the generated power on the other hand [1]. This part of the chapter addresses the potent approaches for increasing volumetric power density. The leading solutions for cutting the volume are using the novel electrodes with reduced thickness and decreasing the channel depth. Also, some of the key remedies for enhancing power are channel indentation, using a novel 3D mesh flow field to manage the liquid water not clogging the routes, and using metal foam instead of traditional flow fields. In order to increase the power density of fuel cells in transportation applications, it is important to increase the performance of fuel cells along with reducing weight and volume. This section introduces major methods for increasing power that is not accompanied by a significant increase in weight and volume, along with new methods that lead to weight loss and volume. These methods can focus on fuel cell components, stack, stack cooling, peripherals as well as power generation management. However, the methods discussed in this section include a review of the most important methods for improving bipolar plates in the fuel cell stack.

7.1.2 Bipolar Plate

Figure 7.1 shows a schematic view of a conventional PEMFC. It contains a membrane being sandwiched between anode and cathode electrodes. As shown in this figure, both electrodes contain (i) a machined flow-field pattern within the bipolar plate (BPP), (ii) a gas diffusion layer (GDL), and (iii) a catalyst layer (CL). Bipolar plate (BPP) is a vital component in the fuel cell stack; it supplies fuel and oxidant to reactive sites, removes reaction products, collects produced current, and provides mechanical support for the cells in the stack. Moreover, in PEMFC stacks, the BPP and GDL are significant cost-driver absorbing around 40% of the total stack cost [2] and also occupying over 90% of the total weight and volume of the stack [3]. Therefore, BPP material and flow-field pattern optimization can greatly improve stack performance, weight, volume, and cost.

Different innovative ideas have been considered to improve the bipolar plate of PEMFCs in literature. Here we review some of them with the greatest improvement.

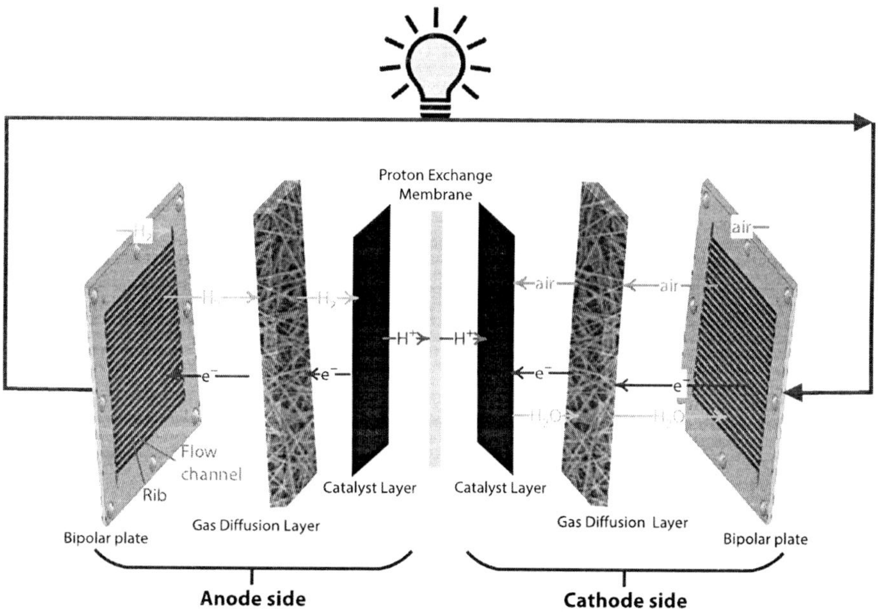

Figure 7.1 A schematic view of a proton exchange membrane fuel cell (PEMFC).

7.1.2.1 Blockages Along the Flow-Field of PEMFCs

This section introduces a method to improve fuel cell performance by inserting blockages into the BPP flow-field channels. This method draws its inspiration from heat transfer improvement techniques employing blockages. Blockages can be considered passive control devices to enhance forced convection heat transfer. According to Guo *et al.* [4] and He and Tao [5], blocks make synergy between convection and conduction terms in heat transfer and increase the heat transfer between the wall and the core flow. This is explained as follows. Consider the energy equation and its advection term, $\vec{V}.\nabla T \equiv u\,\partial T/\partial x + v\,\partial T/\partial y$,

$$\text{Heat-Transfer:} \qquad \underbrace{u\frac{\partial T}{\partial x}+v\frac{\partial T}{\partial x}}_{\equiv\vec{V}.\nabla T}=\alpha\nabla^2 T \tag{7.1}$$

Where $\vec{V}=u\hat{i}+v\hat{j}$ is the local velocity vector, ∇T is the temperature gradient, and α is thermal diffusivity, respectively. In the case of no blocks (Figure 7.2 (a)), ∇T is almost in transverse direction, and $\vec{V}$ is along the channel. Hence the intersection angle between $\vec{V}$ and ∇T is almost 90^0 and $\vec{V}.\nabla T$ is at its minimal value (≈ 0). Placement of blocks within the channel (Figure 7.2 (b)), induces a velocity component in the transverse direction. Hence, the intersection angle between $\vec{V}$ and ∇T will reduce and $\vec{V}.\nabla T$ can, therefore, begin to move away from almost zero value. As a result, convection between the wall and the core flow should also start to increase by the channel blockage. Several papers have investigated the effect of blockages on the improvement of forced convection heat transfer [6, 7]. For example, Heidary and Kermani [7] computed the hydrodynamics and heat transfer enhancement in a channel containing one or more rectangular blocks that partially filled the channel cross-section and reported a 60% improvement in convective heat transfer.

From the analogy between heat and mass transfer phenomena in dynamically similar problems, in a mass transfer problem, it is expected that channel blockage enhances the mass exchange between the channel core part and, say, a catalyst layer at the top boundary of Figure 7.2 (c). Considering the mass concentration equation as:

$$\text{Mass-Transfer:} \qquad \underbrace{u\frac{\partial C_A}{\partial x}+v\frac{\partial C_A}{\partial x}}_{\equiv\vec{V}.\nabla C_A}=D_{A,eff}\nabla^2 C_A \tag{7.2}$$

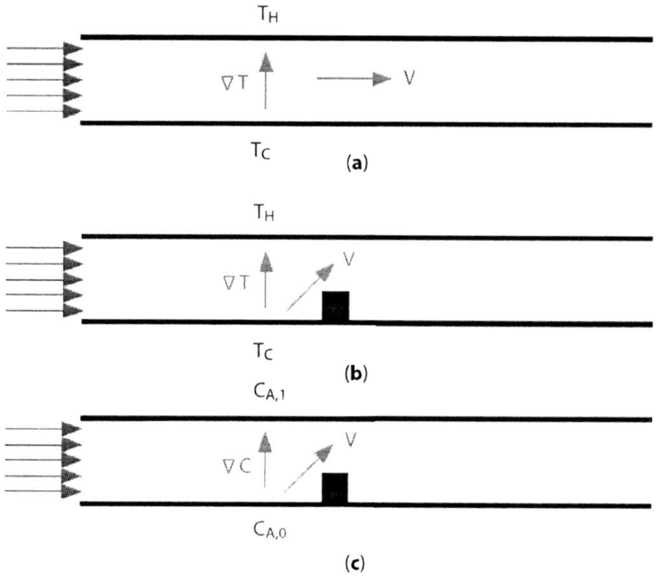

Figure 7.2 The effect of blockages on heat and mass transfer improvement; (a) heat transfer in a straight channel with no block, (b) heat transfer in a straight channel with block, (c) mass transfer in a straight channel with block [8].

Here ∇C_A is the concentration gradient and $D_{A,\,eff}$ is effective binary diffusivity of the reacting species. It is noted that by the blockage of the channels, the intersection angle between $\vec{V}$ and ∇C_A will reduce from 90^0 and $\vec{V}.\nabla C_A$ similar to $\vec{V}.\nabla T$ will start to move from almost zero value. Indeed, such a placement of blocks in PEMFC flow channels can facilitate over-block convection, thereby driving reactant gas convectively into the gas diffusion layer and delivering reactant species directly to the catalyst sites. Over-block convection also aids the removal of reaction products from the gas diffusion layer into the channel, which can further improve performance. Heidary and Kermani [9] have completely analyzed the analogy between heat and mass transfer in a partially porous channel of BPP of a Direct Methanol FC (DMFC) and studied the synergy between different mechanisms of heat and mass transfer. They have shown channel indentation can improve heat transfer along the partially porous channel up to 40% and predicted that the mass transfer between the core channel and the catalyst layer in the fuel cell (FC) would be increased in the same order with blockages [2]. Heidary *et al.* [8] have investigated the effect of block placement along the flow channels of the anode/cathode side of PEMFCs and studied the influence of the blocks number, blocks height, and full or partial blockages. They concluded that anode blockage does not benefit due to lower losses in the anode side of PEMFCs and higher hydrogen diffusivity compared to oxygen. Therefore, they have considered blockage on the only cathode side. They have shown that partial blocks improve cell performance, especially in the area of reduced concentration of the polarization curve. Compared to full blockages, it has less improvement in cell performance, because in the case of partial obstruction, there is a space on the blocks where the species can pass through the flow channel without diffusion in the porous area. But in the case of a complete blockage, the only way left for the gas to pass is through excessive convection of the block through the porous zone (GDL/CL), which improves the performance of the complete blockage. It is observed that the presence of blockage directs the reactive gases into the GDL and increases the molar concentration of oxygen on the catalyst layer, which increases the cell performance, especially in the limited mass transport region of the polarization curve. Their parametric studies showed that the cathode-side channel indentation with five blocks and 100% blockage increases the net power by up to 30% [8]. Some researchers have studied the effect of blocks with different geometries along the flow channels of PEMFCs. Zhang *et al.* [10] have proposed the single-channel PEMFC with novel wedge-shaped fins in cathode-side flow channels. They showed that the outlet oxygen mass fraction is lower by increasing the fin's volume, indicating reactant consumption increase. They also showed that the growth rate of power density is enhanced by increasing the volume or number of wedges. The effect of cross-section shapes of blockages and dimensionless blockage length has been discussed by Cai *et al.* [11]. Via numerical modeling, they have shown that the channel with blockages composed of flow guide parts and suppression parts could increase the performance of PEMFC by 4.4% and 9.3%, respectively. It was also found that trapezoidal cross-sectional indentation with the flow-guide part upstream and the suppression part downstream is 16% higher than straight channels. Ebrahimzadeh *et al.* [12] performed a complete numerical simulation and also experimental evaluations to study the influence of obstacle dimensions and geometry. Considering maximizing species consumption (maximizing current density) and minimizing pressure drop together, they found the optimum dimensions for the obstacle and concluded that the triangular and square shapes are the optimum kinds of blocks which may enhance the current density by 50% with respect to the unobstructed channel [12]. Kuo *et al.* [13] performed numerical modeling to study the performance characteristics of PEM fuel cells with wave-like flow channels. Perng *et al.* [14, 15] also numerically investigated the installation of baffles in the flow channel of PEMFCs and found that the fuel cell performance was enhanced. Wu and Ku [16] developed a three-dimensional model to study PEM fuel cell performance using multiple transversely-inserted rectangular cylinders in flow channels and found higher performance with a reasonable pressure drop. Guo and Qin [17] have investigated two new channels (one-block and two-block) with different blocks to analyze their effects on water removal using the volume of fluid (VOF) model. Although one-block can accelerate liquid water removal, however, their two-block channel design achieves both faster water removal and smaller pressure drop.

In addition to the numerical studies listed above, the literature also contains a few experimental investigations of the effect of block placements in PEMFC flow-field channels. For example, Belcher *et al.* [18] compared the PEMFC performance of parallel-serpentine-baffle and parallel-serpentine flow channels experimentally. They showed that the parallel-serpentine-baffle configuration exhibited better performance under low humidity conditions due to improved water retention in the flow-field channels. Thitakamol *et al.* [19] experimentally studied a mid-baffle interdigitated flow field and concluded that its power output was approximately 1.2–1.3 times greater than the conventional one. Han *et al.* [20] studied the effect of wall waviness of the flow channel on PEMFC performance and showed that concentration loss induced by unstable mass transfer is delayed, and the fuel cell's performance is improved by 5%. Ku and Wu [21] investigated a novel design consisting of rectangular baffles within an interdigitated flow field by simulation and experiment and reported that the in-line baffle pattern increases the rate of electrochemical reaction and net power. The in-line pattern of the blocks in a flow-field is a layout that positions of blocks in two adjacent channels are the same when measured from channel inlet, as shown in Figure 7.3-left. On the contrary, the staggered layouts for the positions of blocks in two adjacent channels are in the zig-zag pattern, as shown in Figure 7.3-right. Shen *et al.* [22] have considered five types of flow field based on a single serpentine flow field, including conventional flow field and in-line blockages with various intervals. They have performed numerical and experimental studies and concluded that with the addition of blockages, fuel cell performance improved with the increased pressure drop, indicating that the pay-to-benefit ratio should be considered during the structure optimization. Fahruddin *et al.* [23] have discussed three types of flow fields, including conventional parallel, leaf, and leaf-baffle on the cathode side of PEM fuel cell and, via experimental testing, shown that the leaf-baffle flow field increased the fuel cell performance by 37.14% compared to parallels. In another paper, Heidary *et al.* [24] experimentally investigate two different configurations of blockages (in-line and staggered) within a parallel PEMFC flow field and compare

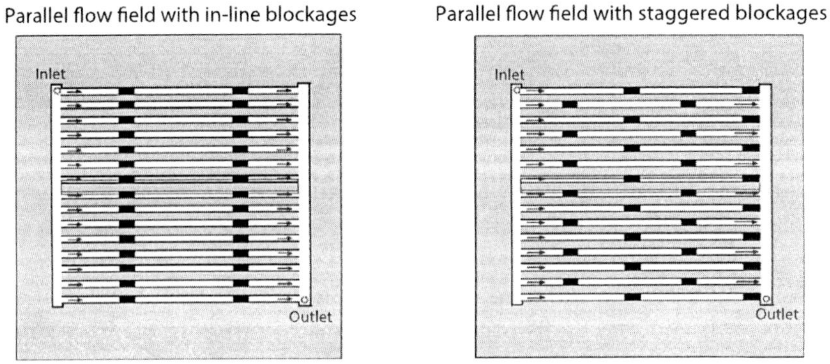

Figure 7.3 Different configurations of blockages along with the parallel flow fields [24].

their electrochemical performance and pressure drop with a baseline parallel flow field without blockages. Figure 7.3 schematically shows the in-line configuration and the staggered configuration. Three blockages were installed in each flow field channel for the in-line and staggered blocked cases. The staggered configuration is superior to the in-line case because the buildup of pressure differentials between adjacent channels drives reactant gas laterally, and the resulting under-rib convection further enhances performance. While the improved performance for the in-line configuration over the baseline case is solely due to under-block convection, the staggered configuration takes advantage of both under-block and under-rib convection (Figure 7.4). Compared to the in-line case, the staggered design shows an 18% increase in the maximum gross power, as well as an 18% reduction in pressure drop at an air flow rate of 1.0 slpm. They have also concluded that fuel cell performance for the staggered case is more stable than both the in-line and baseline cases at high air flow rates. Because of the high flow rates, the presence of blockages helps to retain

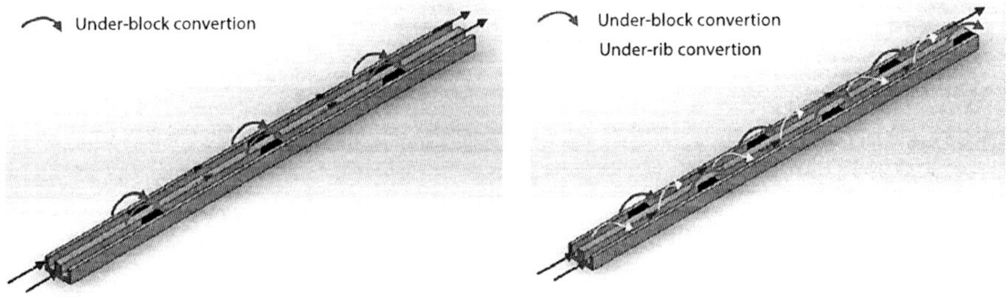

Figure 7.4 Different convection mechanisms for in-line and staggered configurations of blockages along with the parallel flow fields [24].

liquid water within the channels, which enhances membrane hydration and improves its protonic conductivity.

Heidary *et al.* [25] have conducted a numerical study to analyze the under-rib and under-block convection in a parallel flow-field with in-line and staggered blockage configurations. Their numerical simulation was 3D, multi-phase, non-isothermal numerical modeling of a single cell, counter flow, 9-layer PEM fuel cell. Results showed that the staggered configuration enhances maximum net power by up to 11% over the baseline case and by 7% when compared to the in-line case. The presence of under-rib convection in the staggered configuration reduces the pressure drop by 70% compared to the in-line case, which only experiences under-block convection. This research group has performed another study to investigate the effects of channel blocking in a conventional interdigitated flow field of PEMFCs both numerically and experimentally [26]. Their experimented cell contains a 25 cm^2 active area tested at four air flow rates 0.4, 0.7, 1.0, and 1.5 slpm. A 3D simulation of a repeating unit of a whole cell was used for numerical modeling. They have concluded that channel blocking at 1.5 slpm improves the limiting current density by 9% and enhances the maximum net power (the generated power from which the pumping power is subtracted) by 22%. The same idea was used for the parallel flow field of PEMFCs by Wang *et al.* [27]. They have developed a 3D, multiphase PEMFC model as well as an experimental study to investigate flow field design, including the conventional parallel flow field, parallel trapezoid baffle plate, and staggered trapezoid baffle plate flow fields, and concluded that the maximum net power of staggered configuration is nearly 6.4% and 2.5% more than the conventional parallel flow field and parallel trapezoid baffle plate, respectively.

7.1.3 Bio-Inspired Flow Fields

Bio-inspired flow fields are other alternatives presented in the literature to enhance the uniformity of reactant flow distribution and consequently improve the fuel cell performance. In fact, there are different samples of material transport systems such as water, nutrient, oxygen and etc., in plants and animals. The most famous samples are hierarchical structures in leaves of trees or lungs of animals. As an idea, the researchers and manufacturers may benefit from these transport systems to design bio-inspired flow fields for fuel cell systems. For example, Figure 7.5 illustrates different fuel cell flow fields samples designed based on the leaf's hierarchical structure [28]. In this study, Guo *et al.* [28] presented three different bio-inspired flow fields from leaf's structure by considering three generations of channels as

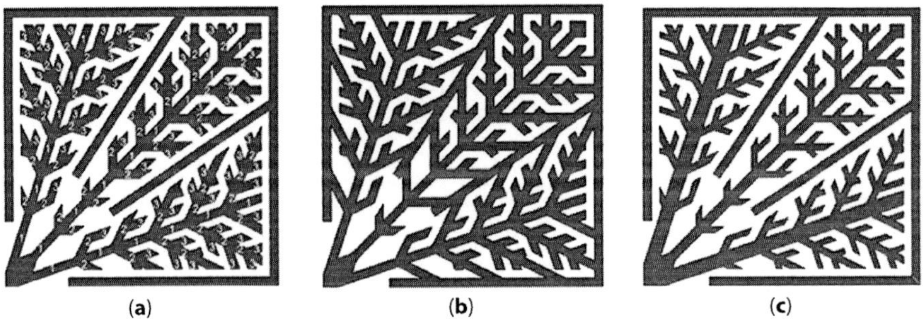

Figure 7.5 (a) Interdigitated flow field with constant channel width, (b) non-interdigitated bio-inspired design with constant, (c) interdigitated design with varying channel width determined by Murray's law [28].

numbered in Figure 7.5 (a) which contains 1) interdigitated flow field with constant channel width (Figure 7.5 (a)), 2) non-interdigitated bio-inspired design with constant (Figure 7.5 (b)), and 3) interdigitated design with varying channel width determined by Murray's law[1] (Figure 7.5 (c)). In the interdigitated designs, the flow field is divided into three regions in which the inlet branches were not directly connected to the outlet channels. The flow must penetrate into the GDL and reach the outlet channels. In the non-interdigitated flow field, the outlet channels are connected to the second-generation channels, and the inlet channels are not connected to the outlets. The results of CFD simulation and experimental study of Guo *et al.* [28] showed that the interdigitated bio-inspired flow fields are more successful in distributing the reactant flow over the GDL and enhancing the maximum power density by 20-25% compared to conventional interdigitated parallel or parallel-series flow field designs. However, as a penalty, pressure drop has been increased in the bio-spired interdigitated flow fields design compared to the non-interdigitated case.

As another example, Badduri *et al.* [29] evaluated the power performance of fuel cells with bio-inspired flow channels based on human lung and leaf design through an interesting experimental investigation. They concluded that interdigitating the leaf channel design enhances the net power (after considering the pressure drops) by 5.58% because of more uniformity in fluid flow with respect to the serpentine flow field.

[1]According to the Murray's law, the dimensions (both length and diameter) of vascular segments generally become smaller from the parent to daughter vessels (from primary to secondary and tertiary generations) in order to minimize the required biological work for transporting nutrients within the plants.

7.1.4 Metal Foam

Porous metallic foam can be used in PEMFCs as a rib-less flow distributor alternative to conventional bipolar plates with rib-channels. Several experimental and numerical studies have investigated the effect of this replacement, and the results show the metallic foam has better performance than traditional BPPs due to being able to deliver a more uniform reactant supply at the face of electrodes, providing higher cell power density [30, 210]; lower ohmic losses and better exchanges of heat between the electrodes and coolant; providing more uniform temperature distribution and removing localized maximum temperatures [31]; reducing the weight and machining costs of stacks w.r.t the graphite plates, i.e., conventional ribbed cells [30, 32]; and lessening the mass transport losses at high current densities [33]. Also, the high porosity of metal foams can help to reduce the pumping powers [30].

Park *et al.* [34] have used graphene foam as the flow distributor of the fuel cell and investigated its effect on increasing the transfer of reactants. The results of their comparison between the conventional rib-channel flow distributor with rib-less graphene foam (GF) flow distributor have been shown in Figure 7.6 [34]. As shown in Figure 7.6 (a), the GF flow distributor could reduce the fuel cell thickness and consequently cut its volume, which means that the GF flow distributor may enhance the volumetric power density (=generated power per cell volume in terms of kW L^{-1}). Based on the results of laboratory tests and comparison of GF and the conventional flow distributor, they concluded that the mass transfer of

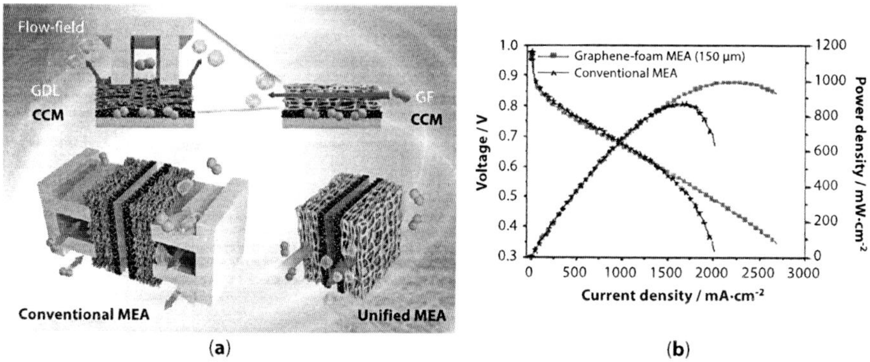

Figure 7.6 (a) Schematically comparison between conventional fuel cell based on rib-channel flow distributor and based on rib-less graphene foam (GF) flow distributor, (b) Polarization curve of PEM fuel cell with conventional flow distributor and rib-less graphene foam (GF) flow distributor [34].

reactants and products is improved and the current density is significantly improved, especially in the area of concentration drop, see Figure 7.6 (b) [34]. On the other hand, the lower thickness of the foam design drains the water produced, which in turn reduces the concentration drop.

For the first time, Park *et al.* [34] have integrated GDL and rib-channel flow field of PEM fuel cells via using graphene foam flow distributor, see Figure 7.7. Removal of GDL in the fuel cell reduces the thickness of the MEA, thereby reducing the reactants pathway from the flow field to the catalyst layer may also cause to decrease in the mass transport limitation. The use of integrated foam also reduces the activation loss by increasing the flow path pressure drop. It also facilitates the evacuation of water droplets by increasing the velocity of reactants due to the thickness of the layers. All of the above factors have led to improved fuel cell performance and current density and power, which is mainly due to reduced activation loss and concentration losses. The results of numerical simulations and experimental tests of Park *et al.* [34] showed that their innovative design has a better performance than the conventional design while reducing the thickness of the layers improves the volumetric power density by up to 82%.

In another study, Park *et al.* [35] implemented the copper foam flow field instead of the serpentine flow field. They inspected the microstructure of various copper foams and investigated their effect as a flow field on PEMFCs. They concluded that PEMFCs with the optimized foam flow field deliver the highest performance reported to date (2.2 A cm^{-2} at 0.6 V). The CO poisoning test showed that the MEA foam flow field showed a higher performance decrease than the normal MEA flow field, which indicates an increase in the use of the reactants [35].

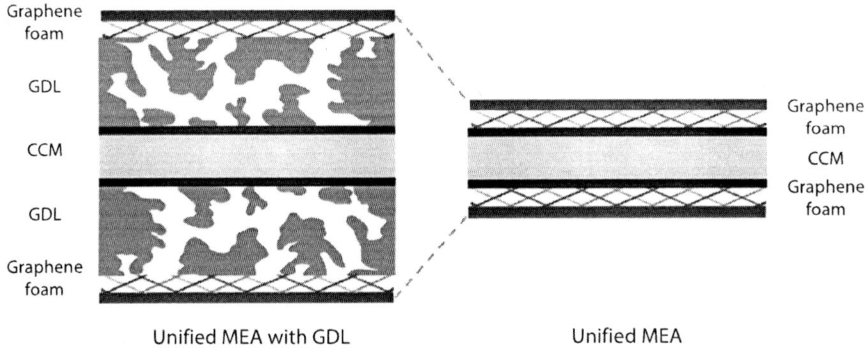

Figure 7.7 Reducing the cell thickness by replacing both the rib-channel and GDL with a graphene foam flow distributor [34].

Based on Kermani *et al.* [1], the main two advantages of substituting conventional flow distributor+GDLs with metal foams are reducing stack volume and increasing power through attaining uniform fluxes of reacting species at the face of the catalyst layer. The foam flow distributor may reduce the stack weight by 40% compared to the conventional graphite bipolar plates (BPPs) [35]. Through an experimental investigation, Awin and Duthan showed that the performance of fuel cells with foam flow distributors is improved by 9.9% with respect to the parallel serpentine flow field [36]. The superiority of foam flow distributors for increasing the power density also have been revealed by Huo *et al.*'s experiments [37]. Carton and Olabi also presented the same results via computational fluid dynamics (CFD) simulation of fuel cells with a double channel flow field and an open-pore cellular foam flow field [38]. The foam-based flow fields can apply a more uniform current density distribution, lower concentration losses, and proper membrane hydration [39]. Zhang *et al.* performed a CFD simulation according to modeling the complete foam structure (morphology) through modeling all ligaments, pores, and nodes and compared it with a conventional serpentine channel fuel cell [40]. Results revealed that the fuel cell with foam flow field does not experience any concentration loss, and its performance is enhanced at high current densities. Due to high porosity (> 90%) and lack of ribs in foam flow distributor, the reactants flow have been distributed more evenly at each section from inlet to outlet compared to the conventional rib-channeled flow fields [41, 42]. The results presented in simulation research in the literature indicate that even with considering the pressure drop, the foam-based flow distributors have the with maximum net power (best trade-off between reactant flow uniformity and pressure drop) compared to other alternatives such as straight parallel, multi-channel serpentine, and new serpentine design [43]. Azarafza *et al.* also reported a 10-50% improvement in maximum power density of fuel cells with foam flow distributor w.r.t. the parallel, serpentine, interdigitated, blocked-parallel flow fields [44]. They showed that the foam flow distributor experiences a higher power density because of lower liquid water saturation, more uniform flow reactant and temperature distributions, and a medium pressure drop. Foam flow distributors enable to accelerate the liquid water removal rate and avoid flooding at high current densities [41], which means postponing the limiting current density [37]. However, the wettability of the foam and GDL layer plays a key role in flooding in these porous media [45]. The liquid water removal rate of foam flow fields may be increased by multi-layered graphene coating on the metal foam [46] or foam with hydrophobic ligament [47]. Bao *et al.* revealed that the liquid water discharge rate from foam flow distributors could be increased by increasing the hydrophobicity of ligaments and reducing the pore size of the foam [48]. Of course, to drain the droplets with the same size of foam pores, requires a stronger convective flow.

In brief, the main pro and cons of foam-based flow distributors which are reported in the literature, are listed as follows:

Pros: i) gas uniformity on the catalyst layer surface [30], ii) mass transport losses alleviation [49], iii) the temperature distribution is also more even for the cells with foam flow distributor than prevalent parallel channel fuel cells [31], iv) the maximum temperature is also lower in cells with foam flow distributor, v) the temperature uniformity of foam flow distributor can keep the membrane at a safe temperature range and prevent it from localized dehydration [36], vi) high porous metal foam could also provide the required membrane hydration more easily under the low relative humidity condition [39].

Cons: i) corrosion [50], ii) contact resistance of metal foam-based electrodes [51].

Corrosion also occurs for the metallic bipolar plate and metal foams, graphite foams [52], or graphene-based foams [34, 53]. So, it is a shared disadvantage for foam-based and conventional flow fields. The remedy is to use anti-corrosion surface coatings such as graphite [52] or PTFE coating [41] on the foam surface.

Contact resistance may be increased at the interfaces of the foam/current collector and foam/catalyst layer because of omitting the solid channel lands (which are an excellent conductor of electrons) and also the difference between the scale of pore sizes in foams and catalyst layer. This contact resistance could be handled by compressing the foams [54] and using a microporous layer (MPL) sandwiched between the foam and catalyst layer [55, 56].

Recently, a few research studies have been presented in the literature that uses metal foams with variable porosities to distribute the reactant flow more evenly on the whole surface of the electrode from inlet to exit. In this way, the uniform flow could remove the low-velocity corner regions and deliver the reactant gas to the whole catalyst layer surface, increasing cell performance. On the other hand, removing the low-velocity corner regions could avoid local starvations and prolong the fuel cell durability. This issue will be described later in section 2-4-2-. In this regard, To enhance the performance of fuel cells with the current homogeneous foams, Shin *et al.* proposed to use a two-piece foam flow field; the first piece had a small pore size and was placed near to the inlet to increase the contact surface area (reducing contact resistance), the second piece has large pore size and placed near to outlet to facilitate gas diffusion [57]. Their proposed variable pore size foam increased the maximum power by 60.1%. In another experimental study, the influence of using variable porosity metal foam (VPMF) as a cathode flow field with five diagonal segments of different porosity has been investigated and tried to conduct the flow from the core region (the main diagonal path from inlet to outlet) into the concave low-velocity corners to apply more uniform flow [58]. VPMF could enhance the maximum power by 8.32% with respect to the homogeneous foams. In the newest research, Kermani *et al.* proposed a novel VPMF that contains different segments with an especial distribution of porosity, permeability, and Forchheimer coefficient [1]. This special distribution increases the flow resistance for the short flow paths (the main core flow region) from inlet to exit and conversely reduces the flow resistance for the long flow paths (which pass through the low-velocity corners). In fact, they fragmented the foam flow distributor into smaller segments and determined the permeability of each segment based on the segment's length, so that $K \propto L$ [1]. So, the shorter segment, the lower the permeability (lower porosity and higher Forchheimer coefficient) and vice versa. Under this situation, the VPMF has been designed so that the flow resistance is the same for all possible paths such as the core flow with a shorter length or the flow passing through the low-velocity corners with a longer length. This novel VPMF could remove these corners and apply a more uniform flow distribution on the catalyst layer surface, enhancing the volumetric power density by 80% compared to conventional parallel serpentine flow fields [1]. Finally, they concluded that this improvement is due to cutting the cell volume and more uniform reactants distribution [1].

7.1.5 Recent Progress in Bipolar Plates of Vehicular Fuel Cells

In recent years, bipolar plates with three-dimensional structures in fuel cells have attracted much attention, especially from vehicle manufacturers. This structure has significant advantages in terms of mass transfer and water management. One of the most important types of bipolar plates is the 3D fine mesh flow field design. Toyota designed the 3D fine-mesh flow field as cathode flow distributor of the Mirai 2014 stack instead of the straight type in the 2008 model stack [59]. The ribs became narrower compared to the 2008 version. This 3D fine-mesh flow field creates two separate paths for liquid water and air, avoiding water pocket accumulation in

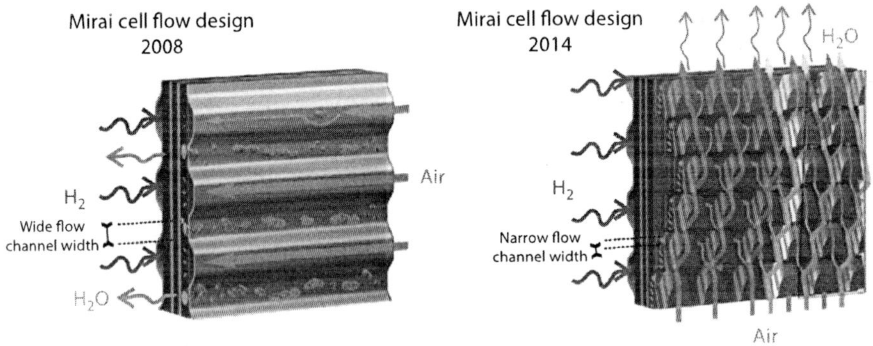

Figure 7.8 Comparing conventional parallel flow field in the fuel cell of Mirai 2008 and the new 3D fine mesh flow field in the fuel cell of Mirai 2014 [60].

the gas channels and conducting the air into the GDL easier than before [60]. This 3D fine-mesh flow field has been illustrated in Figure 7.8 [60].

Bao *et al.* [61] have reconstructed the morphology of this 3D flow field based on an optical microscope image to discuss the single- and two-phase flow characteristics via the VOF model. This 3D fine mesh flow field design has been proposed before by Toyota for Mirai [59]. They have studied the effects of air inlet velocity, droplet size, and baffle contact angle. Based on their results, this 3D fine mesh flow field design has the ability of air guidance, which passes the air through the 3D baffles and guided towards the surface of GDL, which creates a convective flow vertical to the bipolar plate. They also revealed that the liquid in the upper baffle area could be retained, and the water coverage could be reduced. Zhang *et al.* also performed a computational fluid dynamic multi-phase model of a PEM fuel cell equipped with a 3D fine mesh flow field, which was previously designed by Toyota [62]. Their results showed that the 3D fine mesh flow field improves the supply rate of the gas reactants and liquid water removal during the fuel operation. Therefore, the limiting current density increases with no changes in pumping power (meaning that the concentration loss at high current densities is deceased) due to improvement in mass transport and the vertical orientation of the flow. They also revealed that the ohmic loss increment caused by reducing the surface contact of the GDL and the 3D flow field is not significant.

In another attempt, Shen *et al.* designed a new 3D flow field to facilitate the liquid water drainage and increase the FC performance [63]. They modeled a 3D flow field with a straight flow unit and staggered inclined subchannels covered by the hydrophilic coating. Results indicated that the

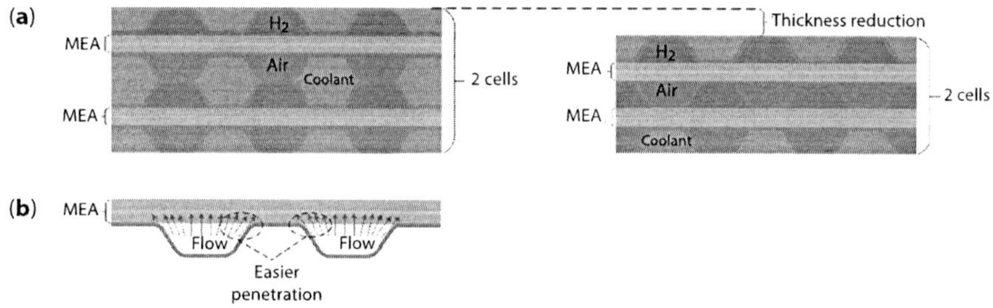

Figure 7.9 Honda Clarity flow field modifications, (a) stack thickness reduction, (b) V-shape channel [169].

3D flow field could easily separate the liquid water from the gas flow and increase the drainage rate. Also, they found that the geometry of this 3D flow field has a synergetic effect of improving the convective terms and the effective mass transfer coefficient. As a result, the mass transport is enhanced, and PEM fuel cell performance is improved.

As another pioneer fuel cell vehicle manufacturer, Honda also modified the bipolar plate configuration of Honda-Clarity compared to the 2006 model's cell. They designed a new V-shape flow channel and succeeded in reducing the stack thickness by about 20% so that every two cells are supported just by one coolant channel with just 1 mm thick, as shown in Figure 7.9(a) [64]. While in the 2006 model, there is one coolant channel for each cell. Therefore, the fuel cell stack of this new configuration is more compact with considerably lower volume and weight (about 33%) [64]. On the other hand, the V-shape flow channel increases the reactant diffusion into electrodes by striking the gas molecules to the oblique walls and taking them away into the catalyst layer, see Figure 7.9(b) [64].

7.2 Fuel Cell Degradation Mechanisms

7.2.1 Introduction

Durability is one of the unsolved challenges for fuel cell systems to commercialize in the transport system. By now, the fuel cells' lifespan is not competitive with that of their traditional competitor, internal combustion engines (ICE). Generally, all aging phenomena can be classified into three categories including, mechanical, thermal, and chemical degradation [65, 66]. Mechanical degradation involves any process that decreases fuel cell components' mechanical properties (e.g., strength reduction, permanent deformation, failure) [66]. Thermal degradation includes sudden thermal shocks, overheat, and operating temperature fluctuations that negatively impact ionomer and membrane and cause performance loss [66, 67]. Chemical degradation refers to the components' almost morphological and structural changes caused during fuel cell operation [65]. Fuel cell vehicular operation leads to chemical degradation. Since the primary goal of this chapter is to investigate automotive fuel cell aging, the mechanical and thermal degradations are out of the scope of the current study. Automotive operation on real roads imposes strict conditions on fuel cells, reducing lifespan. Start-stop cycling, open-circuit voltage (OCV)/idling operation, potential (load) cycling, high power are the most vehicular conditions which a fuel cell may experience as powertrain of a vehicle [68–70]. Figure 7.10 illustrates a schematic of voltage loads applied to the fuel cell under each condition.

These conditions may harm the fuel cell's components, especially the catalyst layer and membrane, decreasing durability. Fuel cell lifespan could be estimated based on the degradation rates of vehicular conditions as follows [68, 71]:

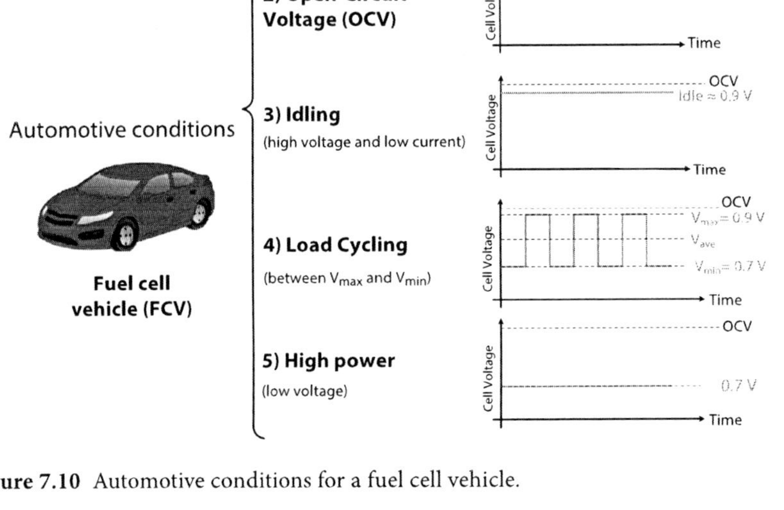

Figure 7.10 Automotive conditions for a fuel cell vehicle.

$$T_f = \frac{\Delta V}{k_p(DR_{ss} \cdot n_{ss} + DR_{id} \cdot t_{id} + DR_{lc} \cdot n_{lc} + DR_{hp} \cdot t_{hp})}. \tag{7.3}$$

In which, ΔV is the maximum allowable voltage drop at the rated power in the whole life of the fuel cell, k_p is a deterioration factor that considers the difference between the real and tentative driving cycles on roads and in the laboratory, respectively. *DR* is the degradation rate of each vehicular condition, including start-stop cycling (DR_{ss}), idling (DR_{id}), load cycling (D_{lc}), and high power (DR_{hp}). n_{ss} and n_{lc} are the cycle numbers of start-stop and load cycling per hour of the driving cycle, respectively. t_{id} and t_{hp} are the time of idling and high-power operations per hour of the driving cycle, respectively. Based on the Department of Energy (DOE) in the United States, the maximum allowable voltage drop after an accelerated stress protocol (AST) is considered 10% [72, 73]. DOE has set initial and ultimate targets of automotive fuel cell lifespan to be 5000 and 8000 hours, respectively, to be competitive with conventional ICEs [72]. The Chinese government has targeted 6000 hours for the lifespan of passenger fuel cell vehicles in the near future by 2025, and 8000 hours as the final target by 2030 [74, 75]. However, the current durability is about 4000-4500 hours [68] and is still far from meeting the ultimate targets. Although there is not a deep comprehension about the reasons for fuel cell degradation, and it may exist more details in this regard compared to what has been presented in the literature, however, the most significant and famous phenomena and chemical changes happen during vehicular conditions and leading to reduce the fuel cell performance, and lifespan are studied and introduced here, such as the platinum (Pt) particles coarsening by dissolution and redeposition (Ostwald ripening) [76, 77], Pt particle migration and coalescence [78, 79], Pt-band formation [79, 80], Pt particle detachment [81], free radical formation [82], carbon corrosion [83, 84], hot spot by gas crossover [85–87], ionomer redistribution [88, 89], etc. Recognition of these phenomena could help to find a solution to control and mitigate the degradation rates and enhance the fuel cell lifespan. Therefore, the operation of each automotive condition and its degradation mechanisms are investigated.

Investigation on long-term durability and lifespan of fuel cells under actual vehicular conditions of real roads is very time-consuming and expensive, and their results will usually be available very later than the fuel cell technology progress rate [75, 90]. Hence, different accelerated stress tests (ASTs) are designed to simulate the actual conditions for the fuel cell and help estimate the lifespan with a considerably shorter time and lower cost. Up to now, lots of ASTs have been presented in the literature to determine the fuel cell lifespan [75]. However, it is tough and almost impossible to compare their results with each other because the fuel cell has been tested under various operating conditions [75] (temperature, pressure, relative humidity, inlet flow), different ASTs' properties (static, dynamic, amplitude and frequency, wave shape). Also, the reference current density in which the performance degradation rate has been determined differs. In fact, because of various climate, air quality, environment, people driving habits, traffic management systems around the world, so different ASTs have been developed for pioneer countries such as Canada, China, EN, South Korea, US, etc. based on the regional requirements of the vehicle [75]. Many investigations compared different AST protocols for fuel cell lifespan estimation [91, 92]. Bloom *et al.* compared different AST protocols, including DST (suggested by the US Department of Energy); the New European Drive Cycle (ECE R15); IEC-TS62282-7-1 (IEC), and FCTT (presented by the US Driving Research and Innovation for Vehicle Efficiency and Energy Sustainability Fuel Cell Technical Team) [93]. Chen *et al.* [75] and Zhou *et al.* [94] also investigated and compared different AST protocols which universities and research institutes have proposed in Canada, China, EU, India, Japan, South Korea, US. Despite all the progress performed in computing fuel cells lifespan by ASTs, the lack of a standard and worldwide AST protocol used to estimate the durability of fuel cell systems for different vehicle segments (light to heavy-duty vehicles) is felt. A definite AST protocol with unit operating conditions and properties is very helpful to compute and compare different fuel cells' lifespans with each other.

The degradation phenomena are classified into two groups, reversible and irreversible. Usually, both reversible and irreversible degradation processes occur during fuel cell operation. Reversible degradation, which contains the more significant part of the whole degradation [95], is usually caused by constant fuel cell operation without changes for a long time [96]. Performance loss due to water flooding [97], ionomer dehydration, Pt oxidation and dissolution during constant operation at OCV/idling [98], some kind of gas poisoning (e.g., CO poisoning) are examples of reversible degradations. It may be recovered and retrieved by applying some recovery procedures, including interrupting the constant operation to cease the OCV/idling [99], gas purging for liquid water removal in flooding case [96], reducing cathode potential to avoid Pt oxidation at OCV/idling [96] and potential cycling [100]. Based on the experimental results of Zago *et al.*, most of the performance loss caused by Pt oxidation at high potential operation could be recovered just by applying a low potential operation for a short time [98]. However, Gazdzick *et al.* suggested a long shutdown process for 15 h as a recovery action between the AST [96]. During this shutting down, no current density is drawn from the fuel cell, gas reactant pumping is stopped, the cell is cooled down to ambient temperature, and finally, the cell voltage will vanish as soon as the residual reactants consume. Based on Gazdzick *et al.*'s investigation, CO contaminant at the anode side may be recovered by applying operation potential higher than 0.6 V to the anode electrode which is available during the shutdown process (it will be explained later in section 2-2-) [96]. The cathode potential reduction to lower than 0.65 V during shutdown also decreases Pt oxidation coverage [96]. On the other hand, although irreversible degradation comprises a little part of degradation [95], its changes are permanent and unrecoverable. Here, all the focus is on the irreversible degradations, introducing their process, and studying their mitigation process.

In conclusion, it seems that it is required to use some recovery interruptions between the ASTs, to eliminate the effects of reversible degradations on the aging measurements of a fuel cell.

7.2.2 Start-Stop Cycling

Start-up and shutdown cycles are one of the AST protocols and significant conditions experienced by a fuel cell vehicle 38, 500 cycles per its lifespan [65], which equals 33% of the whole degradation under actual conditions [101, 102]. During the start-up or shutdown, the anode electrode is divided into two zones. The first one is occupied by hydrogen. The cell acts as a regular fuel cell in which the generated protons are transferred through polymer electrolyte membrane (PEM) from the anode to cathode. In contrast, there is air (oxygen), and the cell acts as an electrolyzer in which a reverse protonic current is applied from the cathode to the anode at the second zone. Under this situation, the local cell potential may increase as high as 1.5 V at the second zone and causes severe damages such as carbon corrosion [102].

Since the phenomena during both start-up and stopping processes in the fuel cells are similar, just the details of the start-up process have been described here. As long as the fuel cell is not used and is shut down, both the anode and cathode electrodes are occupied by air (oxygen); thus, the potentials of both anode and cathode are equal to the potential of the oxygen reduction reaction ($V_a = V_c = 1.23$ V vs. the reference hydrogen electrode (RHE)). Therefore, the net fuel cell voltage, which is the difference between the electrode's potentials ($V_{cell} = V_c - V_a$), is zero (the fuel cell is off).

As soon as the fuel cell is started, the hydrogen flows through the anode electrode and divides the fuel cell into two zones, as shown in Figure 7.11. At zone A which is closer to the anode inlet, the anode catalyst layer (ACL) is exposed by hydrogen, and the hydrogen oxidation reaction takes place at zone A within the anode, which leads to producing protons and electrons and reducing the anode potential to zero (vs. RHE). The protons would be transferred through the PEM from the anode to the cathode (forward protonic current), as shown in Figure 7.11. The cell acts as a fuel cell at zone A, and the net cell potential equals the cathode potential ($V_{cell} = V_c$). So, considering the cathode irreversibility, the cell voltage will reduce to, for example, 0.85 V (as shown in Figure 7.11). On the other hand, at zone B, which is farther from the anode inlet, the hydrogen does not arrive yet, and it is still occupied by air (oxygen). As a result, the oxygen reduction reaction (ORR) occurs at zone B inside the ACL, while at the corresponding zone at the cathode catalyst layer (CCL), the carbon atoms, which are the support base of Pt catalyst, is oxidized by water molecules through carbon oxidation reaction (COR) and produce CO_2, protons, and electrons. Also, the water is decomposed into oxygen, protons, and electrons through an

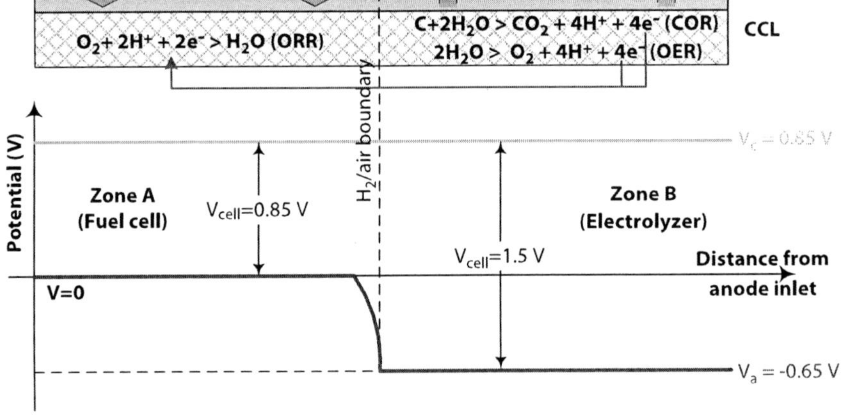

Figure 7.11 The electrochemical reactions during start-stop cycles and the potential distribution along the length of the fuel cell from inlet to outlet of the anode.

oxygen evolution reaction (OER), see Figure 7.11. Under this situation, contrary to the normal routine, the generated protons at the cathode side will be transferred through the PEM to the anode side and lead to a reverse protonic current at zone B. Reiser *et al.* were the first who described the air/hydrogen interface within the ACL and reverse current through the PEM using one-dimensional potential model [103]. According to the ORR at zone B within the anode side, the anode potential may be reduced to say -0.65 V (vs. RHE). So, the net cell voltage would reach 1.5 V ($V_{cell} = V_c - V_a = 0.85-(-0.65)$). As shown in Figure 7.11, the generated electrons at zone A of ACL and zone B of CCL would be transferred to the other zone of that catalyst layers through the electron pathways. While the protons would not be transferred in-plane between zones A and B because the in-plane protonic conductivity of regular membrane is negligible [103]. The shutting-down process is almost similar to the start-up; just the locations of zones A and B are replaced.

Imposing high cell potential (around 1.5 V) can intensify the carbon corrosion rate through COR. Carbon support will be oxidized at high potential and converted to CO and CO_2 [104]. COR is rarely possible at usual fuel cell operation due to its low kinetic [105]. Cell potential over 1.2 V could exacerbate the COR [83, 84]. Shao *et al.* concluded that the rate of carbon oxidation reaction with a potential of 0.207 V is low in normal conditions, but it will be fast at the high potential condition imposed by air/hydrogen interface during start-stop cycles [106]. Unfortunately, the presence of Pt catalyst particles has a negative impact and increases the carbon corrosion rate [107, 108].

Other destructive consequences of carbon corrosion await the catalyst layer, including hole creation [109], shrinking carbon support [81] and decreasing catalyst layer thickness [110], Pt degradations [81, 105, 111], ionomer redistribution [89]. Carbon corrosion weakens the catalyst layer and may cause pinholes in some cases [109]. Ishigami *et al.* simulated actual start-stop conditions by exchanging the anode feed gas between H2 and air [109]. After 500 cycles, their scanning transmission electron microscope (TEM) image revealed that oversized pinholes were created at the inlet and outlet regions inside the cathode catalyst layer. Because opposite the middle area of the fuel cell, the regions near the inlet and outlet are exposed to reverse current and high voltage in start-stop cycles. In addition to collapsing the carbon support and thinning the catalyst layer, these pinholes may block the pores of catalyst layers which are the reactant routes to reach the Pt surface [112]. Schulenburg *et al.* studied the effect of carbon corrosion of reducing the catalyst layer porosity by visualization of 3D pores of catalyst layer [112]. So, carbon corrosion may also lead to increasing the mass transfer resistance [112]. Lin *et al.* reported that the carbon support corrosion reduced the thickness of the cathode catalyst layer by 60% after 1800 start-stop cycles [113]. Yu *et al.* reported a 75% reduction for catalyst layer thickness after 1500 start-stop cycles [114].

Carbon corrosion caused by start-stop cycles weakens the carbon support and detaches the Pt particles, and consequently dissolution and aggregation of Pt particles [102]. These issues reduce the electrochemical surface area (ECSA) and, finally, performance decay [102]. Ettingshausen *et al.* studied the effect of start-stop cycles on Pt dissolution and agglomeration using a transmission electron microscope [105]. The dissolution rate is enhanced by frequent oxidation and reduction of Pt particles (due to cycling between start-up and normal conditions). They found that Pt particles grew as the carbon supports shrank [105]. Pt particles tended to aggregate and ruin the porous properties of the catalyst layer. Qui *et al.* investigated the effect of start-stop cycling with relative humidity (RH) cycles on the fuel cell morphology and performance [115]. They found that after 40 cycles of start-stop + 300 RH cycles, the Pt particles coarsened by 75%, and ECSA reduced by 48%. Qui *et al.* noted that the high potential condition at cathode electrode during start-stop cycles might lead to aggregate the Pt particles, carbon corrosion, and even cause holes and cracks in the catalyst layer [115].

One of the other side effects of carbon corrosion caused by start-stop cycles is ionomer redistribution [88, 89]. Actually, the morphological changes followed by carbon corrosion cause ionomer agglomeration and redistribution [89]. A portion of Pt particles would be uncovered or overcovered by ionomer, leading to omitting the triple-phase zones (TPZs) [88]. This issue directly affects the ECSA loss and cell performance reduction. It is worth noting that the Pt particles need access to a triple-phase zone to perform the electrochemical reactions on their surface [116]. The surface of each active Pt particle must be covered by three phases: i) solid carbon support for electron transport, ii) ionomer for proton transport, and iii) void pores for reactant gas accessibility. This issue has been illustrated schematically in Figure 7.12. If any of the active particles are totally uncovered or overcovered by the ionomer phase, then the TPZ will vanish, and that Pt particle would become an inactive catalyst particle, which reduces the ECSA and fuel cell performance, consequently. In addition to ECSA loss, the ionomer agglomeration and redistribution may decrease the dissolved water content and play a negative role in the water removal process from the catalyst layer, which means that the water flooding possibility will increase.

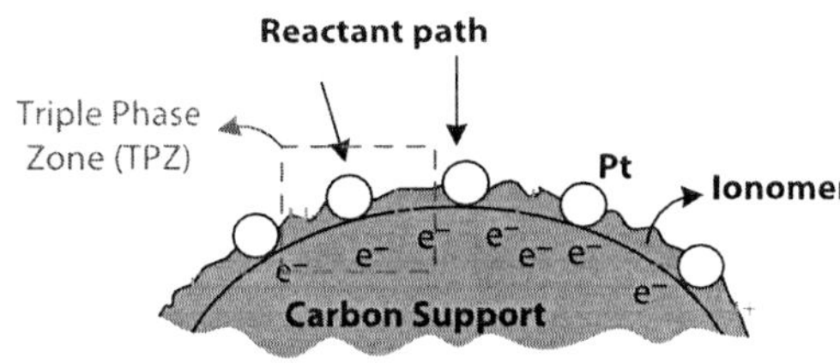

Figure 7.12 Schematic of the triple-phase zone (TPZ) on the surface of Pt particle.

Due to carbon oxidation caused by start-stop cycles, local hydrophilicity may occur at the location of carbon corrosion [117]. This issue accumulates liquid water inside the catalyst layer and is prone to water flooding. As a result, the corroded carbon will increase the mass transport resistance [117]. Park *et al.* evaluated the fuel cell performance at 1.3 V AST, which is near to start-stop cycles with high potential [118]. They showed that the mass transport resistance increased by liquid water accumulation due to the increasing surface hydrophilicity of the corroded carbon.

7.2.3 Open Circuit Voltage (OCV)/Idling Operation

Open circuit voltage (OCV) operation is defined as operation at the highest potential, say 0.95 V, without drawing current density. This condition usually happens exactly after start-up and before shutting down the fuel cell. Idling operation is also defined as the operation at high potential, say 0.9 V, in which a little current density (say 8 [119]-10 [120] mA cm^{-2}) is drawn from the fuel cell to launch the vehicle. A fuel cell vehicle is exposed to OCV and idling conditions about 100 h and 1000 h during its lifespan, respectively [65]. The schematics of OCV and idling loads have been indicated in Figure 7.10. Based on Figure 7.10, both OCV and idling are a kind of static loading at high potential conditions. They are very similar; therefore, their aging phenomena and processes are almost the same. Thus, these two load conditions are introduced together.

7.2.3.1 H_2O_2 Generation and Free Radicals' Attack

OCV/idling operation may underlie some of the aging phenomena such as membrane dehydration, producing hydrogen peroxide (H_2O_2) [120], free radicals (such as hydrogen •H, hydroxyl •OH, hydroperoxyl •OOH) [121], growth of Pt particles due to Ostwald ripening [122], Pt dissolution [76] and migration [123], Pt-band formation [124] which can degrade the fuel cell performance. It is tried to describe these aging phenomena to present a deep comprehension of the degradation reasons to find solutions for avoiding or mitigating them.

OCV/idling operation corresponds to no or a small water production from the electrochemical reaction due to the very low current density. As a result, the polymer electrolyte membrane (PEM) is prone to dehydrate and consequently reduces the protonic conductivity of PEM [125, 126]. This issue will mitigate cell performance.

Due to the meager consumption rates (or no consumption) of the reactant gases during OCV/idling conditions, partial pressure of reactant gases is significant. This high partial pressure may cause a crossover of the reactants through the PEM [95, 127], especially for the oxygen, which has a considerable residual concentration. On the other hand, since the water production is also deficient in this condition, the membrane pores are not clogged by liquid water and are completely open. So, the oxygen molecules can crossover the PEM more easily [128]. Every crossover oxygen molecule reduces with two generated protons and electrons and produces hydrogen peroxide (without electricity generation); its electrochemical reaction is as follows:

$$O_2 + 2H^+ + 2e^- \rightarrow H_2O_2, \quad E_0 = 0.695\ V\ (\text{vs RHE}) \tag{7.4}$$

The electrochemical reaction of Eq. (7.2) takes place at voltage 0.695 V. The potential level of the cathode electrode is very high (0.9-0.95 V) at OCV/idling condition so that the H_2O_2 generation may cease at the cathode side. Due to the lower potential of the anode catalyst layer, it is very prone to generate H_2O_2 by crossover oxygen from the cathode electrode. H_2O_2 generation during OCV/idling conditions has previously been referred by earlier studies [129, 130]. On the other hand, there are always transition metal ions, M^{2+}, such as Fe^{2+}, Cu^{2+}, Co^{2+}, within the fuel cell, especially for metallic bipolar plate (BPP), which are caused during the manufacturing process of the fuel cell or maybe corrosion of the metallic BPP [131]. These transition metal ions, M^{2+}, could react with H_2O_2 to produce free radicals including •H, •OH, •OOH. They are called Fenton's reaction, which is summarized as follows [132]:

$$M^{2+} + H_2O_2 \rightarrow M^{3+} + \bullet\,\text{OH} + \text{OH}^-, \tag{7.5}$$

The new metal ion, M^{3+}, in the product of Eq. (7.5), reacts with H_2O_2 and generates •OOH:

$$M^{3+} + H_2O_2 \rightarrow M^{2+} + \bullet \text{OOH} + \text{OH}^+, \tag{7.6}$$

Other free radicals may be produced through different reactions as follows [133]:

$$H_2O_2 + \bullet \text{OH} \rightarrow \bullet \text{OOH} + \text{H}_2O,$$

$$H_2 + \bullet \text{OH} \rightarrow \bullet \text{H} + \text{H}_2O \tag{7.7}$$

$$O_2 + \bullet \text{H} \rightarrow \bullet \text{OOH}$$

Hydroxyl radical (•OH) could be produced by direct reaction of crossover oxygen from the cathode to anode and the hydrogen as follows:

$$O_2 + H_2 \rightarrow 2 \bullet \text{OH}, \tag{7.8}$$

The presence of free radicals produced in Equations (7.5)-(7.8) may destroy the PEM structure and degrades fuel cell performance. Actually, the structure of the most common membranes used in fuel cells is made up of Perfluorosulfonic acid (PFSA), which consists of three parts [134]: 1) a polymeric backbone (main chain) which is Polytetrafluoroethylene (PTFE), 2) a vinyl ether side chain, and 3) a polar sulphonic acid group (HSO_3^-) which terminates the side chain. This is an ionic polymer which is call ionomer, and its structural formula, including these three parts, is shown in Figure 7.13. Since the sulphonic acid group is a polar terminal, it will be attacked by the generated free radicals (•H, •OH, •OOH). An attack of free radicals may degrade some of the weak bonds such as α-OCF_2, β-OCF_2, and C-S bonds [135, 136]. This issue leads to reducing membrane thickness [135], increasing leakage current [137], and adverse effects of the protonic conductivity [135], which may be measured by the fluoride emission rate (FER) in the outlet water streams of the fuel cell. The free radical attacks may finally cause thinning of the membrane and make pinholes and cracks. Under this condition, gas crossover facilitates through the membrane [137]. As a result, free radicals play a destructive role and attack the membrane structure. The production rate of metal ions and free radicals is more severe in the high potential condition such as OCV/idling. For example, Wong's experiments showed that the Fe^{2+} concentration would be decreased by 93% as the cell voltage reduced from OCV to 0.7 V [121].

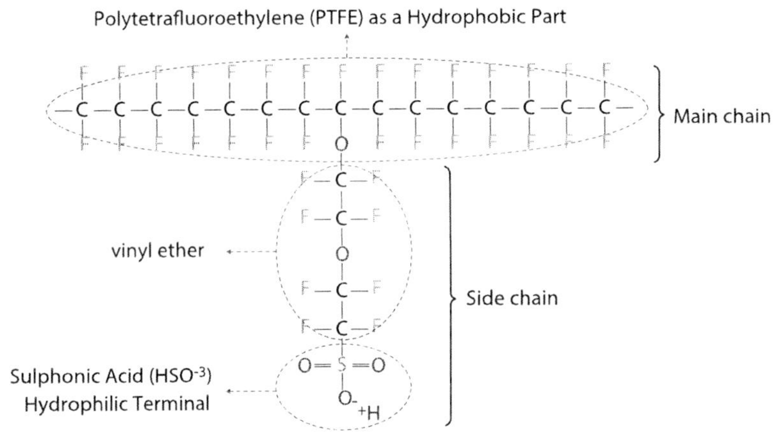

Figure 7.13 The structural formula of PFSA ionomer.

The chemical degradation of the membrane by the attack of free radicals produced during OCV/idling operation has been investigated in the literature. Singh *et al.* modeled the hydroxyl radical attack to the Nafion membrane's ether terminal through four steps sub-reactions [138]. Their results could predict the degradation rate of the experimental test. Ghassemzadeh *et al.* revealed that the hydroxyl radical just attack the side chain of PFSA ionomer, which contains vinyl ether and sulphonic acid group at its terminal [135], and the main chain (PTFE) is safe from the radical attacks. •OH radical can attack the terminal of the side chain, sulphonic acid group, by weakening both the C-S bond and C-O bonds [129, 139]. In another study, Ghassemzadeh *et al.* investigated the formation and reactivity of hydrogen and hydroxyl radicals with Nafion 211 as the membrane [136]. They showed that the hydrogen radicals could attack the tertiary carbon in both the main and side chains while the hydroxyl radical solely destroys the side chain structure. Some previous research expressed the sequence of chemical degradation reactions of the membrane caused by free radical attacks in more detail [129, 138]. Gubler *et al.* simulated the formation of free radicals •H, •OH, •OOH according to Fenton reactions based on H_2O_2 and Fe^{2+} as transitional metal ions [140]. Their results showed that the oxidative strength of free radicals might be ranked as •OH > •H > •OOH.

7.2.3.2 Pt Catalyst Degradation

The automotive conditions such as OCV/idling and load cycling may intensify the Pt catalyst particles degradation through coarsening, detaching, and Pt-band formation. These side effects will be described in the current section.

Pt particles' growth and agglomeration reduce the ECSA and cell performance [211]. The most reasons for Pt sintering are dissolution-redeposition (Ostwald ripening) and migration-coalescence [80]. The former usually occurs at a high potential, such as OCV/idling operation [141], and the latter happens at low potential [78]. So, the Ostwald ripening has been described here, and the coalescence mechanism is deferred to section 2-5-.

Ostwald Ripening Phenomenon: in Ostwald ripening, small and pristine Pt particles placed on the carbon support surface may lose two electrons via their conductive supports and become Pt^{2+} ions. The Pt^{2+} ions have the capability of dissolution within the adjacent ionomer. The dissolved Pt^{2+} ions aggregated with each other to make larger particles. The aggregated Pt particles will precipitate (redeposit) on the carbon support surface with receiving the two electrons again. The Pt agglomeration process through the Ostwald ripening is illustrated in Figure 7.14(a). The small Pt particles are coarsened during this process and converted to large Pt ones by two following steps:

$$\text{Dissolution:} \qquad \text{Small } Pt \rightarrow Pt^{2+} + 2e^{-}, \tag{7.9}$$

$$\text{Precipitation (redeposition):} \quad Pt^{2+} + 2e^{-} \rightarrow \text{Large } Pt, \tag{7.10}$$

Pt particles will grow from small to large ones through the sequence dissolution and redeposition of the Ostwald ripening [76, 142]. Figure 7.14(b) shows the Ostwald ripening timeline. During this time, the small Pt particles precipitate and form larger particles. It seems that the small Pt particles shrink and sinter on the larger ones, and the large particles grow by Pt agglomeration [143]. One can say the large particles swallow the small ones [144]. Based on Gilbert *et al.*'s experiments, the Pt particles with a radius less than the specific radius shrink, and vice versa, Pt particles grow with a radius larger than that [145]. In this regard, the average dimensions of platinum particles increase, but their number decreases [141], as shown in Figure 7.14(b). The main reason for Ostwald ripening is reducing interfacial energy [141]. Moein-Jahromi *et al.* proved that the interfacial energy of two small Pt particles would be reduced by aggregating to each other through Ostwald ripening [141]. They derived a mathematical model to simulate the Ostwald ripening of Pt particles based on the theory

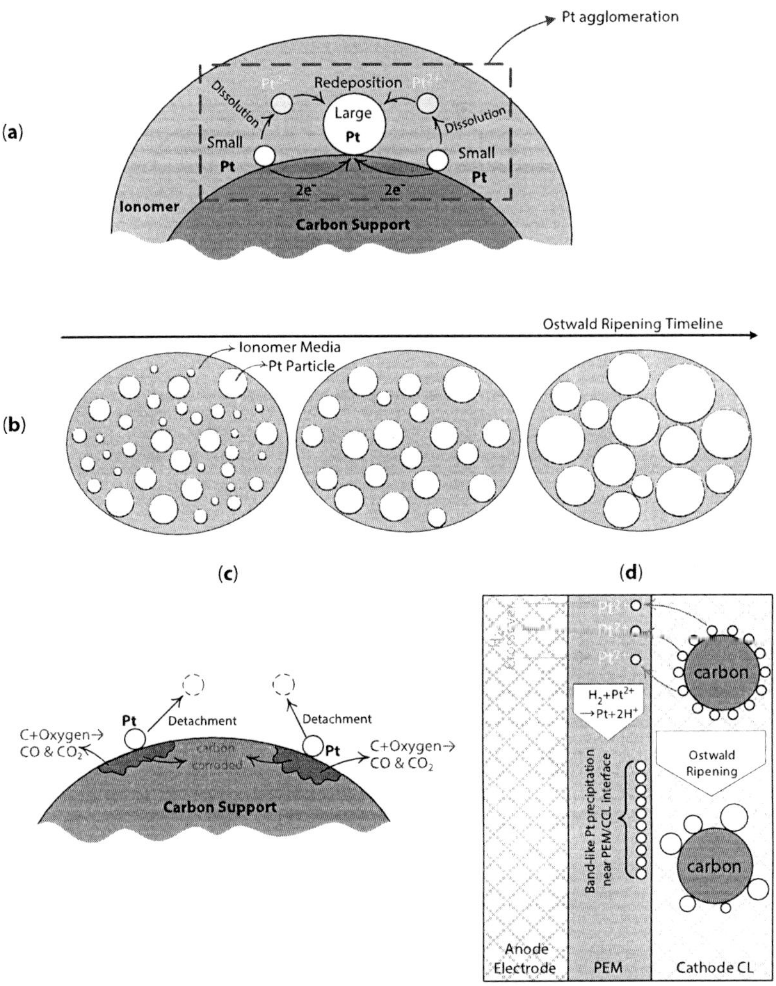

Figure 7.14 (a) Ostwald ripening mechanism, (b) Ostwald ripening timeline, (c) Pt particle detachment, (d) Pt-band formation near the PEM/CCL.

developed by Lifshitz, Slyozov [146], and Wagner [147] (LSW) for dilute dissolutions. The time variation of Pt particle radius is obtained as [141]:

$$R_{pt}^3(t) - R_{pt}^3(0) = \frac{8\gamma_{Pt} C_{\infty,Pt} \Omega_{Pt}^2 D_{Pt,i}}{9RT}.t \tag{7.11}$$

$R_{Pt}(t)$ and $R_{Pt}(0)$ are the average Pt radius of the degraded and pristine particles, respectively. γ_{Pt}, $C_{\infty,Pt}$, Ω_{Pt}, $D_{Pt,i}$ are interfacial energy surface density, solute concentration of flat particle with infinity radius, molar volume, diffusion coefficient in ionomer phase for the Pt particles, respectively.

Some other alternatives can make Pt^{2+} ions and Pt dissolution within the ionomer phase. In addition to Eq. (7.9), Pt particles also may be oxidated to form PtO, and then PtO will be reacted with protons to produce Pt^{2+} ions as follows [122, 148]:

$$\text{Oxidation: } Pt + H_2O \rightarrow PtO + 2H^+ + 2e^-, \tag{7.12}$$

$$\text{Pt Oxide dissolution: } PtO + 2H^+ \rightarrow Pt^{2+} + H_2O, \tag{7.13}$$

Summation of Eqs. (7.12) and (7.13), yields to Eq. (7.9). Eqs. (7.12) and (7.13) just show the other way of Pt^{2+} dissolution, which is the first step of the Ostwald ripening process.

It is worth noting that high potential conditions during the OCV/idling operation intensify the Pt particle solubility into the ionomer phase [78, 80, 141]. Although the Pt solubility dependency on potential is not entirely realized, the results of Ferreira *et al.*'s study showed that the Pt^{2+} equilibrium concentration is logarithmically proportional to the cathode potential [80]. So, any increase in cell potential leads to a considerable rise in Pt^{2+} equilibrium concentration. Ferreira *et al.* reported a voltage decay rate of 25 $\mu V\ h^{-1}$ for the stack OCV operation (0.95 V vs. RHE) during 2000 h [80]. So, OCV/idling may exacerbate the dissolution rate of Pt^{2+} and the Ostwald ripening rate.

Catalyst Detachment: another degradation process is Pt particles detachment from the carbon support [149], leading to Pt loading loss, ECSA, and consequently the fuel cell performance. As mentioned earlier, operation at high potential during OCV/idling conditions causes carbon oxidation (carbon corrosion) [112]. Therefore, local carbon corrosion may occur on the carbon support surface, as indicated in Figure 7.14(c). So, Pt particles may detach from the corroded local place of the carbon support and dissolve into the ionomer. The detached Pt would be dissolved in the ionomer phase and recrystallized (buried) in the membrane [150]. Due to the higher electrode potential, the concentration of dissolved Pt ions is even higher in the cathode electrode compared to the anode side [80]. Hence, the dissolved Pt ions would be diffused from the cathode with higher concentration to the anode through the membrane based on Fick's law. It is the dissolution and migration phenomena of Pt particles that yield to lay within the membrane [76, 151]. It means that the active Pt particles with the triple-phase zone (i.e., ECSA) would be missed and lost. These Pt particles are considered inactive catalysts which cannot participate in the electrochemical reactions.

Pt-band formation: another side effect of Pt dissolution and migration is Pt-band formation near the PEM/CCL interface [152]. Scanning electron microscope (SEM) images from actual aged samples of MEA during OCV/idling conditions reveal a Pt-band next to the PEM/CCL interface [80, 152]. In fact, due to the gradient of Pt dissolution concentration at both sides of PEM, the dissolved Pt particles tend to migrate from the cathode side with higher concentration to the anode side. Therefore, Pt ions would pass through the membrane based on Fick's law. On the other hand, the crossover hydrogen molecules from the anode to the cathode may meet the dissolved Pt^{2+} ions (which migrate in the opposite direction) near the PEM/CCL interface. The dissolved Pt^{2+} ions would react by the crossover hydrogen flow as follows:

$$Pt^{2+} + H_2 \rightarrow Pt + 2H^+, \tag{7.14}$$

The produced Pt atoms from Eq. (7.14) are precipitated and sintered near to the PEM/CCL interface and form a Pt-band [123]. Figure 7.14(d) indicates the schematic of Pt-band formation. Figure 7.14(d) also illustrates the growth of Pt particles via Ostwald ripening over time inside the CCL. The Pt-band location depends on the partial pressure of the reactants at the cathode and anode sides. The Pt-band will be distanced from the PEM/CCL if the hydrogen partial pressure of the anode decreases or oxygen partial pressure of the cathode is increased [123].

7.2.4 Load Cycling

This type of loading, common in automotive applications of a fuel cell, involves the oscillation in extracted power between the OCV/idling (i.e., 0.9 V vs. RHE) and the rated power (i.e., 0.7 V vs. RHE), see Figure 7.10. Load cycling may also underlie temperature and relative humidity fluctuation during the fuel cell operation. The load cycling is considered up to 300, 000 cycles during a fuel cell vehicle lifespan [65]. So, it is the most prevalent automotive condition exposed to a fuel cell vehicle [70]. Load cycling will cause reversible degradation that may decay the performance, such as water flooding [153] and membrane degradation due to hydration-dehydration cycles [154, 155]. Also, some irreversible degradation such as ECSA reduction due to Ostwald ripening, migration and precipitation, detachment of Pt particles may be occurred. Some of the aging side effects caused by load cycling will be discussed here.

7.2.4.1 Mechanical Degradation of Load Cycling

The variation of electrochemical reaction rates induced by load cycling leads to cause a fluctuation in water production. So, it directly impacts the amount of liquid water fraction and water content of the ionomer. This issue leads to a continuous change in the protonic conductivity of the membrane and will raise the ohmic resistance [156] and reduce the membrane's durability. The experimental results revealed that the hydration-dehydration cycles induced because of load cycling operation causes to apply a swelling-shrinking cycle on the membrane volume, which eventually leads to creating pinholes [157], cracks [158], and delamination [159, 160] in the membrane, say mechanical membrane degradation. These membrane pinholes would increase the H_2 crossover; for example, Liu *et al.* measured that the gas crossover current may be increased as high as 17 mA cm^{-2} during the load cycling [161]. On the other hand, the strain rates of catalyst layers and membranes which are in contact with each other are not the same. Therefore, the deformation of the catalyst layers and the membrane is different with regard to relative humidity fluctuations. This issue applies residual stress and strain between the layers and finally leads to delamination [159, 162]. The electronic/protonic conduction will be reduced because of delamination, which causes an increase in the contact resistance between different layers and consequently mitigates cell performance. The other side effect of load cycling is ionomer redistribution inside the catalyst layer due to the relative humidity cycles [163]. As mentioned in sections 2-2-, ionomer redistribution may cause to uncover or over cover the active Pt particles and destroy the required triple-phase zones (TPZ) for the electrochemical reaction [88]. In a nutshell, one of the effects of load cycling is mechanical degradations caused to change the morphology of components and structural damages (including cracking, pinholes, delamination, and ionomer redistribution). There are also different reports in the literature for catalyst layer cracks and delamination during load cycling operation [164–167].

7.2.4.2 Starvation

Starvation is another phenomenon that is possible during load cycling conditions. Actually, starvation may occur when the reactants' supply rates are not as high as the rate of external load change of the fuel cell. Usually, there might be a lag between the rate of load increment during load cycling and reactants flow rates enhancement which leads to oxidant/fuel starvation at the cathode/anode electrode [168]. Unfortunately, most of the prevalent flow fields used within the anode and cathode electrodes could not distribute the reactants on the whole catalyst layer surface uniformly [1, 169]. Generally, the concentrations of reactants are high at the inlet regions and over the channels. The regions over the ribs and near the corners isolated from the main flow path suffer from low reactant concentration [1]. The non-uniform distribution caused by the flow fields would intensify the starvation caused by load cycling. Here both air and fuel starvations and their effects are described.

- **Air starvation:** during load cycling and lack of quick response from the reactant flow rate to the increment in demand load, air starvation occurs in the cathode electrode [168]. In this situation, the required oxidant is not reachable for the transferred protons from the anode side to complete the oxygen reduction reaction (ORR). So, the protons inevitably react with electrons transferred from the anode through hydrogen evolution reaction (HER), $2H^+ 2e^- \rightarrow H_2$, to produce H_2 gas. Experimental observations verify the presence of hydrogen at the combination of outlet cathode gas during air starvation [170, 171]. So, the cathode electrode is prone to be a bed for the direct reaction between the produced hydrogen and the low available oxygen, which leads to local hot spots [172]. Local hot spots exacerbate Pt particles sintering and agglomeration, leading to the aging catalyst layer [173].
- **Fuel starvation:** lack of required hydrogen supply during a sudden load increment cause fuel starvation at the anode electrode. Fuel starvation is more severe in dead-end electrodes than recirculating ones [174]. Fuel starvation during load cycling will decrease the local hydrogen partial pressure at the anode side; as a result, the air molecules can permeate more quickly through the membrane and reach the anode side. When a part of the anode is occupied by hydrogen, and the other part is engaged with oxygen, carbon oxidation reaction (COR) and oxygen evolution reaction (OER) will occur, leading to corroding the carbon supports. As mentioned earlier in section 2-2-, the simultaneous presence of

hydrogen and oxygen within the anode catalyst layer creates a reverse current (see Figure 7.11). Reverse current in anode intensifies the carbon corrosion, Pt agglomeration, catalyst layer thinning, and in a word, accelerates catalyst layers degradation [175]. If the fuel stoichiometry is lower than one and the fuel cell is exposed to extract extra loading, complete fuel starvation happens within the whole anode electrode, then OER and COR are taken place at the anode catalyst layer to produce the required protons, which is called cell reversal [176]. Evidence of this is the presence of oxygen and carbon dioxide in the exhaust gas at the anode side during fuel starvation experiments [177]. In fact, the fuel cell operates as an electrolyzer, and the complete fuel starvation leads to haste the carbon corrosion rate.

7.2.4.3 Chemical Degradation of Load Cycling

Chemical degradation of the catalyst layers, especially the cathode one [70, 178], is another aging aspect that would be happened by load cycling condition. Load cycling exposes a severe operating condition that exacerbates the growth of Pt particles via dissolution and redeposition (Ostwald ripening) [179, 180], migration and precipitation (coalescence) [78], Pt detachment [181], Pt-band formation near the PEM/CCL interface [182]. These aging mechanisms have been described earlier for OCV/idling operation, so they have not been explained again to avoid repetition. Just some of the experimental observations and empirical-based simulation of the literature will be expressed. Kinoshita *et al.* were the first who investigate the degradation of a phosphoric acid fuel cell under load cycling in 1973 [183]. They showed that the ECSA had been reduced by 70% after 3500 cycles of potential load cycling between 0.5-1.25 V with a 60 s period [183]. Patterson was the first who study the PEM fuel cell aging under load cycling [184]. He reported a 50% reduction in ECSA of the catalyst layer under the load cycling 0.87-1.2 V with a 60 s period. He found that the main reason for aging is Pt detachment from catalyst layer, dissolution within ionomer, and micrometer-scale diffusion toward the membrane to form Pt-band at PEM/CCL interface [78]. Wang *et al.* studied the degradation rate of load cycling between the idle state and the rated power [159]. They found that the degradation rate was increased at the rated power compared to the idle mode. Their results show that the catalyst layer's chemical degradation and the membrane are intensified under rated power conditions. Bi and Fuller also presented their experimental results for PEM fuel cell aging under load cycling 0.87-1.2 V with a 60 s period at different operating temperatures [185]. They found that after 7000 cycles, the ECSA was reduced by 70%, the Pt particles have been growing by 130% with respect to the pristine catalyst layer for the operating temperature of 60°C [185]. The cell voltage also decreased from 0.657 V to 0.558 V at 0.8 A cm^{-2}. Their results also revealed that the higher the operating temperature, the more severe the voltage degradation. Based on the literature, the ECSA decreases logarithmically with respect to the number of load cycles to reach its minimum value [185–187]. Debe *et al.* presented a logarithmic and empirical-based relation for ECSA versus the number of cycles and operating temperature [188]. This relation may estimate the degraded ECSA during load cycling operation. Moein-Jahromi *et al.* inspired this logarithmic relation and developed an empirical-based model to predict the ECSA loss during load cycling using an analogy with carbon steel fatigue [141]. They also modeled the Pt particles Ostwald ripening by Lifshitz, Slyozov and Wagner (LSW) model. Their model could predict the fuel cell performance loss during load cycling. Later, Kneer *et al.* also studied the structural changes in CCL, tuned Moein-Jahromi *et al.*'s model [189], and presented an empirical correlation for the ECSA loss during voltage cycling [190]. They found that the upper potential limit (UPL) of the load cycling and dwell time have the most effective determinative factor on the degradation rate [191]. However, based on Kneer *et al.* [189] and Takei *et al.* [192], the other parameters such as temperature and relative humidity also have impacts on the depredation rate of the fuel cell while the lower potential limit (LPL) of load cycling seems to be not significant. Garcia-Sanchez *et al.* evaluated local degradation under load cycling to find a relationship between performance loss and local cell current density [193]. Based on them, Pt-band formation near PEM/CCL interface is intensified at the regions with higher local current densities. In conclusion, degradation phenomena and estimating PEM fuel cell aging rates under automotive conditions and load cycling as the most probable condition become hot topics to develop fuel cell systems. In this regard, scholars investigate chemical degradations such as Ostwald ripening, Pt migration and precipitation (coalescence), Pt detachment, and band formation to introduce their mechanism to find a solution for postponing them and extending the fuel cell durability.

7.2.5 High Power

Operating above 90% of the rated power [194] with high current density and low voltage, say 0.7 V or lower is the last severe condition that may take place during vehicular loading [68]. Although the high-power

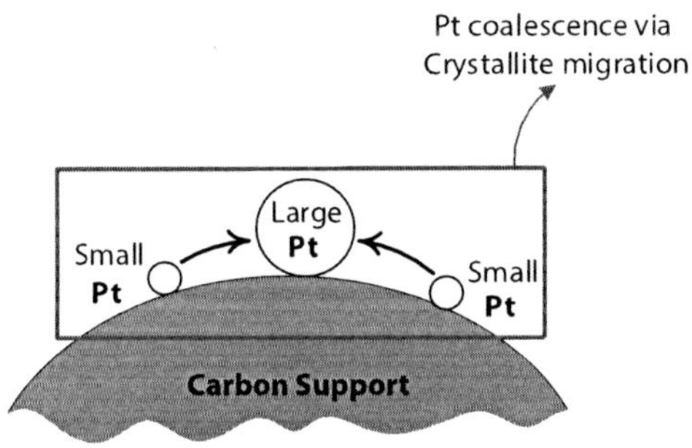

Figure 7.15 Pt coalescence via crystallite migration.

condition has not been investigated widely by the researchers, the most significant side effects of the high-power condition reported in the literature are presented.

Due to the high current extraction during the rated power condition, the electrochemical reaction rate is enhanced, and the liquid water production is raised. So, the fuel cell is prone to water flooding [195]. The reactant pass pores at the gas diffusion layer (GDL), and the catalyst layer may be blocked by liquid water, so the mass transport resistance will be increased [196]. Therefore, the reactant supply will get into trouble at the high current condition; thus, air and fuel starvation might be happened [197]. Starvation will underlie other aging mechanisms, including hot spots [85], uneven reactant distribution, carbon corrosion, Pt particle coarsening, agglomeration, detachment, and finally, catalyst layer degradation [75], described in detail in sections 2-3- and 2-4-.

Another Pt particle coarsening mechanism is crystallite migration and precipitation when they meet each other on the carbon support and recrystallize, which is called coalescence [78, 80]. It is illustrated in Figure 7.15. Crystallite migration and coalescence were reported in the absence of electrolyte in gas-phase Pt/C catalyst with a temperature higher than 500°C [78]. However, since the Pt dissolution rate is negligible at low voltage conditions (high power) [135], the Ostwald ripening is out of the question. So, the coalescence mechanism is supposed to be responsible for Pt agglomeration at high current density (low voltage) [78].

7.2.6 Summary of Aging Mechanisms

Actual automotive driving cycles apply severe conditions to the fuel cell and cause serious aging processes, including mechanical, thermal, and chemical degradations. The membrane and catalyst layer are the most vulnerable components. In this section, to have a complete yet concise summary, all the degradation phenomena introduced during the automotive AST (start-stop cycles, OCV/idling, load cycling, and high power) are listed as a flow chart in the one-glance diagram of Figure 7.16.

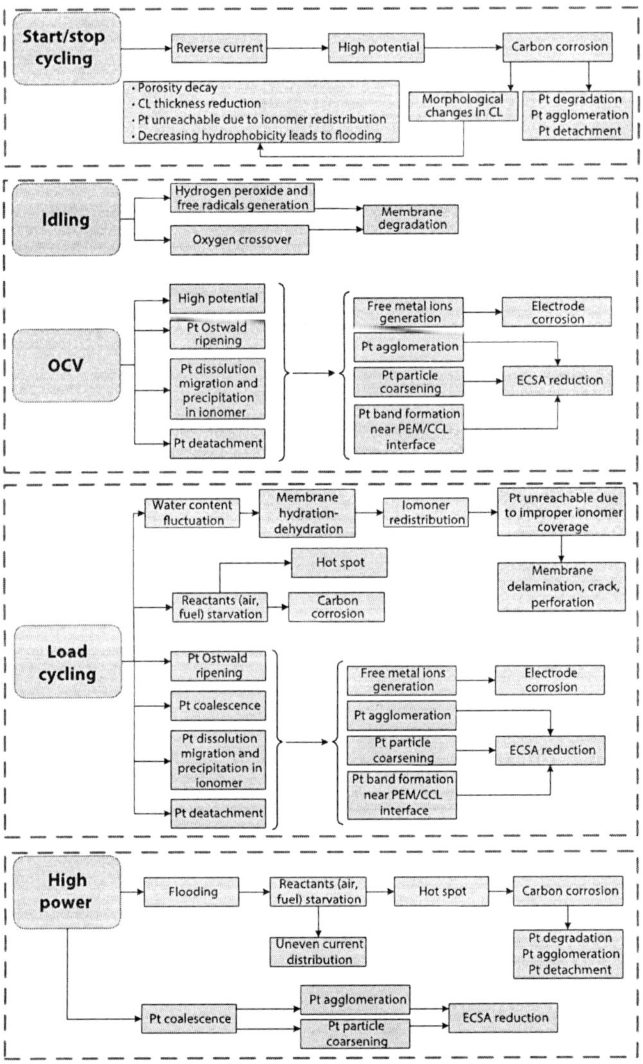

Figure 7.16 Summary of automotive degradation mechanisms.

7.2.7 Measures to Control and Reduce the Degradation Rate of Fuel Cell

As noted earlier, the degradations are divided into two reversible and irreversible groups. In the case of reversible, the suggested recovery procedures include interrupting the constant operation [95] to cease the OCV/idling, gas purging for liquid water removal in flooding case [96], reducing cathode potential to avoid Pt oxidation at OCV/idling [96] and potential cycling [100]. While some different solutions are also presented in the literature for mitigation the irreversible degradation, which are classified into mechanical (e.g., strength reduction, permanent deformation, failure of fuel cell components), thermal (e.g., sudden thermal shocks, overheat, operating temperature fluctuations), and chemical (e.g., morphological and structural changes in the membrane and catalyst layer) categories. Since chemical degradation is the most significant part of the aging mechanism during automotive conditions, the solutions for mitigation of the chemical degradation are presented in this section. Two leading solutions to alleviate the chemical degradation in a fuel cell vehicle are using an energy management strategy (EMS) [68, 198] and improving the material.

- **Energy management strategy (EMS):** EMSs have been developed for electrification vehicles such as fuel cell cars based on different goals such as controlling and reducing degradation rate, improving efficiency and operating near the optimum condition, reducing fuel consumption and operative cost, decreasing greenhouse gas emission [68]. The EMS, run by a power control unit (PCU), decides to divide the external load between the power train sources, battery, and fuel cell stack. Different actions are decided by PCU (based on the EMS) to enhance the fuel cell durability based on the knowledge of vehicular decay phenomena, including reducing the time of OCV/idling operation, the number of start-stop cycles, load cycles, and time of high-power condition [68]. The EMS is categorized into rule-based and optimization-based methods [68]. The rule-based EMS is designed according to the human's deterministic expertise about the EMS cost function. While the optimization-based one is designed according to an optimization cost function, for example, extending the fuel cell durability [68]. As a result, despite the optimization-based method, the role-based is not guaranteed that the EMS forces the fuel cell to operate under its optimum point. For example, Fu *et al.* developed a rule-based EMS according to fuzzy logic for fuel cell/battery/ultracapacitor hybrid vehicles to filter high-frequency load from the fuel cell and the battery to enhance their lifespan [199]. Wang *et al.* presented an optimized-based EMS to minimize two cost functions, the degradation costs as well as the fuel consumption, for a hybrid fuel cell/battery vehicle [200]. They found a middle state in the hybridization range from the battery-dominant to the fuel cell-dominant to mitigate both the degradation rate and the fuel consumption cost [200].

One practical suggestion to prolong the fuel cell lifetime is to use multiple stacks with lower-rated power compared to a single stack with high power. Based on simulation results of Moein-Jahromi *et al.* and Zhang *et al.* [68, 201], the demand power for a fuel cell vehicle could be covered by just one stack out of a triple stack configuration about 82% of the driving cycle, and the two other stacks would be off or in standby mode to decrease the usage time of fuel cell and also reduce the number of start-stop cycles. Therefore, they suggested a triple stack configuration in which the rated power of each stack is one-third of the required single stack [68, 201]. They also defined a specific EMS to split the demand power between these three stacks to minimize the operating time and start-stop cycles [68, 201]. Besides lifetime, the mileage cost of triple stack configuration is also reduced compared to the single stack. In another attempt, Marx *et al.* presented an EMS for a hybrid vehicle of multi-stacks and battery to prolong the fuel cell and battery lifetime based on some experience such as reducing the operating time, operating with a minimum number of the stack(s), limiting the battery discharge depth [202]. Their results revealed that the fuel cell operating time could be reduced by increasing the hybridization degree [202].

- **Material and process:** in order to slow down the rate of catalyst layer degradation in cyclic loading, various methods have been proposed and tested so far. One of these methods is to add a stabilizer metal (mainly noble metals) as a catalyst to the platinum particles. Selvaganesh *et al.* added gold to the Pt/C-based catalyst layer and compared it with the regular catalyst layer [203]. Their results proved that the Pt-Au/C catalyst layer is more resistant to load cycling. In another research, they suggested using RuO_2 to stabilize the catalyst layer against the ASTs [204]. Their results showed that the ECSA reduction and performance loss of regular Pt/C catalyst layer after 7000 load cycles were measured as 50% and 59%, respectively, while these values were just 13% and 7% for Pt-RuO_2/C catalyst layer [204]. Yu *et al.* [205] and Jung *et al.* [206] investigated the influence of using cobalt to stabilize Pt particles. Ostwald ripening and catalyst dissolution were considerably decreased in Pt-Co/C, the potential loss of polarization curve was reduced the ECSA loss is alleviated by 20% [205] or 40% [206] for two different load cycling protocols. Niobium oxide (Nb_2O_5) and Tungsten (W) were also suggested as catalyst stabilizers [207]. In a separate study, Jang *et al.* showed that using some heat treatment, such as annealing the catalyst layer, can significantly contribute to its stability and resistance to load cycling operation [208]. Another solution was proposed by Debe *et al.* to increase the durability of the catalyst layer [188]. They manufactured a kind of nanostructured thin film (NSTF) catalyst, which is decorated with lower Pt mass loading and more resistance under load cycling operation [188]. Novikova *et al.* synthesized catalyst layer based on carbon nanotubes (CNT) and revealed that the CNT-based catalysts are much more resistant to load cycling than conventional catalysts, so that during the load cycling up to 10, 000 cycles, the ECSA loss rate for CNT catalyst layers were 21-58% of that of conventional catalyst [209].

References

1. Kermani, M.J., Moein-Jahromi, M., Hasheminasab, M.R., Wei, L., Guo, J., Jiang, F.M., Development of a variable-porosity metal-foam model for the next fuel cells flow distributors. *Int. J. Hydrogen Energy*, 47, 7, 4772–92, 2022.
2. Heidary, H. and Kermani, M.J., Performance enhancement of fuel cells using bipolar plate duct indentations. *Int. J. Hydrogen Energy*, 38, 5485–96, 2013.
3. Zamora-Antuñano, M.A., Orozco Pimentel, P.E., Orozco-Gamboa, G., García-García, R., Olivárez-Ramírez, J.M., Reyes Santos, E. *et al.*, Flow analysis based on cathodic current using different designs of channel distribution in PEM fuel cells. *Appl. Sci.*, 9, 3615, 2019.

4. Guo, Z.Y., Li, D.Y., Wang, B.X., A novel concept for convective heat transfer enhancement. *Int. J. Heat Mass Transfer*, 41, 2221–5, 1998.
5. He, Y.-L. and Tao, W.-Q., Convective heat transfer enhancement: Mechanisms, techniques, and performance evaluation. *Adv. Heat Transf.*, 46, 87–186, 2014, Elsevier.
6. Heidary, H. and Kermani, M.J., Enhancement of heat exchange in a wavy channel linked to a porous domain; A possible duct geometry for fuel cells. *Int. Commun. Heat Mass Transfer*, 39, 112–20, 2012.
7. Heidary, H. and Kermani, M.J., Heat transfer enhancement in a channel with block (s) effect and utilizing nano-fluid. *Int. J. Therm. Sci.*, 57, 163–71, 2012.
8. Heidary, H., Kermani, M.J., Dabir, B., Influences of bipolar plate channel blockages on PEM fuel cell performances. *Energy Convers. Manage.*, 124, 51–60, 2016, https://doi.org/10.1016/j.enconman.2016.06.043.
9. Heidary, H., Abbassi, A., Kermani, M.J., Enhanced heat transfer with corrugated flow channel in anode side of direct methanol fuel cells. *Energy Convers. Manage.*, 75, 748–60, 2013.
10. Zhang, S., Qu, Z., Xu, H., Talkhoncheh, F.-K., Liu, S., Gao, Q., A numerical study on the performance of PEMFC with wedge-shaped fins in the cathode channel. *Int. J. Hydrogen Energy*, 46, 27700–8, 2021.
11. Cai, Y., Wu, D., Sun, J., Chen, B., The effect of cathode channel blockages on the enhanced mass transfer and performance of PEMFC. *Energy*, 222, 119951, 2021.
12. Ebrahimzadeh, A.A., Khazaee, I., Fasihfar, A., Experimental and numerical investigation of obstacle effect on the performance of PEM fuel cell. *Int. J. Heat Mass Transfer*, 141, 891–904, 2019, https://doi.org/10.1016/j.ijheatmasstransfer.2019.07.034.
13. Kuo, J.-K. and Yen, T.-H., Three-dimensional numerical analysis of PEM fuel cells with straight and wave-like gas flow fields channels. *J. Power Sources*, 177, 96–103, 2008.
14. Perng, S.-W. and Wu, H.-W., A three-dimensional numerical investigation of trapezoid baffles effect on non-isothermal reactant transport and cell net power in a PEMFC. *Appl. Energy*, 143, 81–95, 2015.
15. Perng, S.-W., Wu, H.-W., Shih, G.-J., Effect of prominent gas diffusion layer (GDL) on non-isothermal transport characteristics and cell performance of a proton exchange membrane fuel cell (PEMFC). *Energy*, 88, 126–38, 2015.
16. Wu, H.-W. and Ku, H.-W., The optimal parameters estimation for rectangular cylinders installed transversely in the flow channel of PEMFC from a three-dimensional PEMFC model and the Taguchi method. *Appl. Energy*, 88, 4879–90, 2011.
17. Guo, Q. and Qin, Y., Numerical investigation of water droplet removal characteristics in novel block channels of PEMFC using dynamic wettability model. *Int. J. Hydrogen Energy*, 46, 36890–902, 2021.

18. Belchor, P.M., Forte, M.M.C., Carpenter, D.E.O.S., Parallel serpentine-baffle flow field design for water management in a proton exchange membrane fuel cell. *Int. J. Hydrogen Energy*, 37, 11904–11, 2012.
19. Thitakamol, V., Therdthianwong, A., Therdthianwong, S., Mid-baffle interdigitated flow fields for proton exchange membrane fuel cells. *Int. J. Hydrogen Energy*, 36, 3614–22, 2011.
20. Han, S.-H., Choi, N.-H., Choi, Y.-D., Simulation and experimental analysis on the performance of PEM fuel cell by the wave-like surface design at the cathode channel. *Int. J. Hydrogen Energy*, 39, 2628–38, 2014.
21. Ku, H.-W. and Wu, H.-W., Influences of operational factors on proton exchange membrane fuel cell performance with modified interdigitated flow field design. *J. Power Sources*, 232, 199–208, 2013.
22. Shen, J., Tu, Z., Chan, S.H., Enhancement of mass transfer in a proton exchange membrane fuel cell with blockage in the flow channel. *Appl. Therm. Eng.*, 149, 1408–18, 2019.
23. Fahruddin, A., Ichsani, D., Taufany, F., Widodo, B.U.K., Widodo, W.A., Improved Performance of polymer electrolyte membrane fuel cell using leaf-baffle flow field design. *Int. J. Ambient Energy*, 1–22, 2022.
24. Heidary, H., Kermani, M.J., Advani, S.G., Prasad, A.K., Experimental investigation of in-line and staggered blockages in parallel flowfield channels of PEM fuel cells. *Int. J. Hydrogen Energy*, 41, 6885–93, 2016.
25. Heidary, H., Kermani, M.J., Prasad, A.K., Advani, S.G., Dabir, B., Numerical modelling of in-line and staggered blockages in parallel flowfield channels of PEM fuel cells. *Int. J. Hydrogen Energy*, 42, 2265–77, 2017, https://doi.org/10.1016/j.ijhydene.2016.10.076.
26. Heidary, H., Kermani, M.J., Prasad, A.K., Advani, S.G., Experimental and numerical studies of enhanced interdigitated flow field for PEM fuel cells. *J. Energy Eng.*, 147, 4021026, 2021.
27. Wang, X., Qin, Y., Wu, S., Shangguan, X., Zhang, J., Yin, Y., Numerical and experimental investigation of baffle plate arrangement on proton exchange membrane fuel cell performance. *J. Power Sources*, 457, 228034, 2020, https://doi.org/10.1016/j.jpowsour.2020.228034.
28. Guo, N., Leu, M.C., Koylu, U.O., Bio-inspired flow field designs for polymer electrolyte membrane fuel cells. *Int. J. Hydrogen Energy*, 39, 21185–95, 2014.
29. Badduri, S.R., Srinivasulu, G.N., Rao, S.S., Influence of bio-inspired flow channel designs on the performance of a PEM fuel cell. *Chin. J. Chem. Eng.*, 28, 824–31, 2020, https://doi.org/10.1016/j.cjche.2019.07.010.
30. Zhang, Y., Tao, Y., Shao, J., Application of porous materials for the flow field in polymer electrolyte membrane fuel cells. *J. Power Sources*, 492, 229664, 2021, https://doi.org/10.1016/j.jpowsour.2021.229664.
31. Afshari, E., Computational analysis of heat transfer in a PEM fuel cell with metal foam as a flow field. *J. Therm. Anal. Calorim.*, 139, 2423–34, 2020, https://doi.org/10.1007/s10973-019-08354-x.

32. Wilberforce, T., El Hassan, Z., Ogungbemi, E., Ijaodola, O., Khatib, F.N., Durrant, A. *et al.*, A comprehensive study of the effect of bipolar plate (BP) geometry design on the performance of proton exchange membrane (PEM) fuel cells. *Renewable Sustainable Energy Rev.*, 111, 236–60, 2019, https://doi.org/10.1016/j.rser.2019.04.081.
33. Afshari, E., Mosharaf-Dehkordi, M., Rajabian, H., An investigation of the PEM fuel cells performance with partially restricted cathode flow channels and metal foam as a flow distributor. *Energy*, 118, 705–15, 2017, https://doi.org/10.1016/j.energy.2016.10.101.
34. Park, J.E., Lim, J., Kim, S., Choi, I., Ahn, C.-Y., Hwang, W. *et al.*, Enhancement of mass transport in fuel cells using three-dimensional graphene foam as flow field. *Electrochim. Acta*, 265, 488–96, 2018, https://doi.org/10.1016/j.electacta.2018.01.191.
35. Awin, Y. and Dukhan, N., Metal-foam bipolar plate for PEM fuel cells: Simulations and preliminary results. *Mater. Sci. Forum*, 933, 342–50, 2018, Trans Tech Publ, https://doi.org/10.4028/www.scientific.net/MSF.933.342.
36. Awin, Y. and Dukhan, N., Experimental performance assessment of metal-foam flow fields for proton exchange membrane fuel cells. *Appl. Energy*, 252, 113458, 2019, https://doi.org/10.1016/j.apenergy.2019.113458.
37. Huo, S., Cooper, N.J., Smith, T.L., Park, J.W., Jiao, K., Experimental investigation on PEM fuel cell cold start behavior containing porous metal foam as cathode flow distributor. *Appl. Energy*, 203, 101–14, 2017, https://doi.org/10.1016/j.apenergy.2017.06.028.
38. Carton, J.G. and Olabi, A.-G., Three-dimensional proton exchange membrane fuel cell model: Comparison of double channel and open pore cellular foam flow plates. *Energy*, 136, 185–95, 2017, https://doi.org/10.1016/j.energy.2016.02.010.
39. Jo, A. and Ju, H., Numerical study on applicability of metal foam as flow distributor in polymer electrolyte fuel cells (PEFCs). *Int. J. Hydrogen Energy*, 43, 14012–26, 2018, https://doi.org/10.1016/j.ijhydene.2018.01.003.
40. Zhang, G., Bao, Z., Xie, B., Wang, Y., Jiao, K., Three-dimensional multiphase simulation of PEM fuel cell considering the full morphology of metal foam flow field. *Int. J. Hydrogen Energy*, 46, 2978–89, 2021, https://doi.org/10.1016/j.ijhydene.2020.05.263.
41. Huo, S., Shi, W., Wang, R., Lu, B., Wang, Y., Jiao, K. *et al.*, Elucidating the operating behavior of PEM fuel cell with nickel foam as cathode flow field. *Sci. China Technol. Sci.*, 64, 1041–56, 2021, https://doi.org/10.1007/s11431-020-1767-5.
42. Wan, Z., Sun, Y., Yang, C., Kong, X., Yan, H., Chen, X. *et al.*, Experimental performance investigation on the arrangement of metal foam as flow distributors in proton exchange membrane fuel cell. *Energy Convers. Manage.*, 231, 113846, 2021, https://doi.org/10.1016/j.enconman.2021.113846.
43. Atyabi, S.A., Afshari, E., Zohravi, E., Udemu, C.M., Three-dimensional simulation of different flow fields of proton exchange membrane fuel cell using

a multi-phase, coupled model with cooling channel. *Energy*, 234, 121247, 2021, https://doi.org/10.1016/j.energy.2021.121247.
44. Azarafza, A., Ismail, M.S., Rezakazemi, M., Pourkashanian, M., Comparative study of conventional and unconventional designs of cathode flow fields in PEM fuel cell. *Renewable Sustainable Energy Rev.*, 116, 109420, 2019, https://doi.org/10.1016/j.rser.2019.109420.
45. Jo, A., Ahn, S., Oh, K., Kim, W., Ju, H., Effects of metal foam properties on flow and water distribution in polymer electrolyte fuel cells (PEFCs). *Int. J. Hydrogen Energy*, 43, 14034–46, 2018, https://doi.org/10.1016/j.ijhydene.2018.01.134.
46. Jang, S., Park, H.-Y., Jung, J., Lee, J., Park, H.-Y., Jang, J.H. *et al.*, Enhanced water management of three-dimensional graphene-Ni foam with patterned wettability in a polymer electrolyte membrane fuel cell. *ACS Sustainable Chem. Eng.*, 7, 15487–94, 2019, https://doi.org/10.1021/acssuschemeng.9b03117.
47. Bao, Z., Niu, Z., Jiao, K., Numerical simulation for metal foam two-phase flow field of proton exchange membrane fuel cell. *Int. J. Hydrogen Energy*, 44, 6229–44, 2019, https://doi.org/10.1016/j.ijhydene.2019.01.086.
48. Bao, Z., Wang, Y., Jiao, K., Liquid droplet detachment and dispersion in metal foam flow field of polymer electrolyte membrane fuel cell. *J. Power Sources*, 480, 229150, 2020, https://doi.org/10.1016/j.jpowsour.2020.229150.
49. Zheng, L.J., Srouji, A.K., Turhan, A., Mench, M.M., Computational exploration of ultra-high current PEFC operation with porous flow field. *J. Electrochem. Soc.*, 159, F267, 2012, https://doi.org/10.1149/2.048207jes.
50. Kim, M., Kim, C., Sohn, Y., Application of metal foam as a flow field for PEM fuel cell stack. *Fuel Cells*, 18, 123–8, 2018, https://doi.org/10.1002/fuce.201700180.
51. Kim, J. and Cunningham, N., Development of porous carbon foam polymer electrolyte membrane fuel cell. *J. Power Sources*, 195, 2291–300, 2010, https://doi.org/10.1016/j.jpowsour.2009.10.053.
52. Wu, Y., Cho, J.I.S., Whiteley, M., Rasha, L., Neville, T.P., Ziesche, R. *et al.*, Characterization of water management in metal foam flow-field based polymer electrolyte fuel cells using in-operando neutron radiography. *Int. J. Hydrogen Energy*, 45, 2195–205, 2020, https://doi.org/10.1016/j.ijhydene.2019.11.069.
53. Park, J.E., Lim, J., Lim, M.S., Kim, S., Kim, O.-H., Lee, D.W. *et al.*, Gas diffusion layer/flow-field unified membrane-electrode assembly in fuel cell using graphene foam. *Electrochim. Acta*, 323, 134808, 2019, https://doi.org/10.1016/j.electacta.2019.134808.
54. Liu, R., Zhou, W., Li, S., Li, F., Ling, W., Performance improvement of proton exchange membrane fuel cells with compressed nickel foam as flow field structure. *Int. J. Hydrogen Energy*, 45, 17833–43, 2020, https://doi.org/10.1016/j.ijhydene.2020.04.238.
55. Fultz, D.W. and Chuang, P.-Y.A., The property and performance differences between catalyst coated membrane and catalyst coated diffusion

media. *J. Fuel Cell Sci. Technol.*, 8, 041010 (6 pages), 2011, https://doi.org/10.1115/1.4003632.

56. Sarker, M., Rahman, M.A., Mojica, F., Mehrazi, S., Kort-Kamp, W.J.M., Chuang, P.-Y.A., Experimental and computational study of the microporous layer and hydrophobic treatment in the gas diffusion layer of a proton exchange membrane fuel cell. *J. Power Sources*, 509, 230350, 2021, https://doi.org/10.1016/j.jpowsour.2021.230350.
57. Shin, D.K., Yoo, J.H., Kang, D.G., Kim, M.S., Effect of cell size in metal foam inserted to the air channel of polymer electrolyte membrane fuel cell for high performance. *Renewable Energy*, 115, 663–75, 2018, https://doi.org/10.1016/j.renene.2017.08.085.
58. Kang, D.G., Lee, D.K., Choi, J.M., Shin, D.K., Kim, M.S., Study on the metal foam flow field with porosity gradient in the polymer electrolyte membrane fuel cell. *Renewable Energy*, 156, 931–41, 2020, https://doi.org/10.1016/j.renene.2020.04.142.
59. Toyota, Outline of the Mirai 2016, https://www.toyota-europe.com/download/cms/euen/Toyota Mirai FCV_Posters_LR_tcm-11-564265.pdf, [accessed 15 June 2021].
60. Coralli, A., Sarruf, B.J.M., de Miranda, P.E.V., Osmieri, L., Specchia, S., Minh, N.Q., Chapter 2 - Fuel cells, in: *PEVBT-S and E of H-BET*, de Miranda (Ed.), pp. 39–122, Academic Press, London, UK, 2019, https://doi.org/https://doi.org/10.1016/B978-0-12-814251-6.00002-2.
61. Bao, Z., Niu, Z., Jiao, K., Analysis of single-and two-phase flow characteristics of 3-D fine mesh flow field of proton exchange membrane fuel cells. *J. Power Sources*, 438, 226995, 2019, https://doi.org/10.1016/j.jpowsour.2019.226995.
62. Zhang, G., Xie, B., Bao, Z., Niu, Z., Jiao, K., Multi-phase simulation of proton exchange membrane fuel cell with 3D fine mesh flow field. *Int. J. Energy Res.*, 42, 4697–709, 2018, https://doi.org/10.1002/er.4215.
63. Shen, J., Tu, Z., Chan, S.H., Performance enhancement in a proton exchange membrane fuel cell with a novel 3D flow field. *Appl. Therm. Eng.*, 164, 114464, 2020, https://doi.org/10.1016/j.applthermaleng.2019.114464.
64. Honda Co., Clarity fuel cell powertrain 2020, https://global.honda/innovation/FuelCell/Clarity-Fuel-Cell-engineer-talk.html, [accessed 10 February 2021].
65. Garche, J., Dyer, C.K., Moseley, P.T., Ogumi, Z., Rand, D.A.J., Scrosati, B., *Encyclopedia of electrochemical power sources*, Elsevier, Amsterdam, The Netherlands, 2013.
66. Shen, Y., Mechanical degradation of membrane electrode assemblies in proton exchange membrane fuel cells, PhD thesis, 2017.
67. Tavassoli Kakhki, A., Investigating the local membrane degradation mechanisms in PEM fuel cells, MSc thesis, 2014.
68. Moein-Jahromi, M. and Heidary, H., Durability and economics investigations on triple stack configuration and its power management strategy for

fuel cell vehicles. *Int. J. Hydrogen Energy*, 46, 5740–55, 2021, https://doi.org/10.1016/j.ijhydene.2020.11.103.
69. Chen, K., Laghrouche, S., Djerdir, A., Degradation model of proton exchange membrane fuel cell based on a novel hybrid method. *Appl. Energy*, 252, 113439, 2019, https://doi.org/10.1016/j.apenergy.2019.113439.
70. Moein-Jahromi, M. and Kermani, M.J., Three-dimensional multiphase simulation and multi-objective optimization of PEM fuel cells degradation under automotive cyclic loads. *Energy Convers. Manage.*, 231, 113837, 2021, https://doi.org/10.1016/j.enconman.2021.113837.
71. Chen, H., Pei, P., Song, M., Lifetime prediction and the economic lifetime of proton exchange membrane fuel cells. *Appl. Energy*, 142, 154–63, 2015, https://doi.org/10.1016/j.apenergy.2014.12.062.
72. Satyapal, S., *Annual progress report: DOE hydrogen and fuel cells program*, National Renewable Energy Lab.(NREL), Golden, CO (United States), 2018.
73. Wang, Y., Moura, S.J., Advani, S.G., Prasad, A.K., Power management system for a fuel cell/battery hybrid vehicle incorporating fuel cell and battery degradation. *Int. J. Hydrogen Energy*, 44, 8479–92, 2019, https://doi.org/10.1016/j.ijhydene.2019.02.003.
74. SAE-China, Technology roadmap for energy saving and new energy vehicles, 2016.
75. Chen, H., Song, Z., Zhao, X., Zhang, T., Pei, P., Liang, C., A review of durability test protocols of the proton exchange membrane fuel cells for vehicle. *Appl. Energy*, 224, 289–99, 2018.
76. Ioroi, T., Siroma, Z., Yamazaki, S., Yasuda, K., Electrocatalysts for PEM fuel cells. *Adv. Energy Mater.*, 9, 1801284, 2019.
77. Hansen, T.W., DeLaRiva, A.T., Challa, S.R., Datye, A.K., Sintering of catalytic nanoparticles: Particle migration or Ostwald ripening? *Acc. Chem. Res.*, 46, 1720–30, 2013.
78. Shao-Horn, Y., Sheng, W.C., Chen, S., Ferreira, P.J., Holby, E.F., Morgan, D., Instability of supported platinum nanoparticles in low-temperature fuel cells. *Top. Catal.*, 46, 285–305, 2007, https://doi.org/10.1007/s11244-007-9000-0.
79. Marcu, A., Preparation, characterization and degradation investigations of cathode catalysts for automotive PEM fuel cells systems, PhD thesis, 2014.
80. Ferreira, P.J., Shao-Horn, Y., Morgan, D., Makharia, R., Kocha, S., Gasteiger, H.A., Instability of Pt/C electrocatalysts in proton exchange membrane fuel cells: A mechanistic investigation. *J. Electrochem. Soc.*, 152, A2256, 2005.
81. Takao, S., Sekizawa, O., Samjeské, G., Kaneko, T., Higashi, K., Yoshida, Y. *et al.*, Observation of degradation of Pt and carbon support in polymer electrolyte fuel cell using combined nano-X-ray absorption fine structure and transmission electron microscopy techniques. *ACS Appl. Mater. Interfaces*, 10, 27734–44, 2018.
82. Nosaka, Y., Ohtaka, K., Ohguri, N., Nosaka, A.Y., Detection of OH radicals generated in polymer electrolyte membranes of fuel cells. *J. Electrochem. Soc.*, 158, B430, 2011.

83. Macauley, N., Papadias, D.D., Fairweather, J., Spernjak, D., Langlois, D., Ahluwalia, R. *et al.*, Carbon corrosion in PEM fuel cells and the development of accelerated stress tests. *J. Electrochem. Soc.*, 165, F3148, 2018.
84. Darab, M., Barnett, A.O., Lindbergh, G., Thomassen, M.S., Sunde, S., The influence of catalyst layer thickness on the performance and degradation of PEM fuel cell cathodes with constant catalyst loading. *Electrochim. Acta*, 232, 505–16, 2017.
85. Zhang, X., Yang, D., Luo, M., Dong, Z., Load profile based empirical model for the lifetime prediction of an automotive PEM fuel cell. *Int. J. Hydrogen Energy*, 42, 11868–78, 2017.
86. Inaba, M., Kinumoto, T., Kiriake, M., Umebayashi, R., Tasaka, A., Ogumi, Z., Gas crossover and membrane degradation in polymer electrolyte fuel cells. *Electrochim. Acta*, 51, 5746–53, 2006.
87. Pei, P., Wu, Z., Li, Y., Jia, X., Chen, D., Huang, S., Improved methods to measure hydrogen crossover current in proton exchange membrane fuel cell. *Appl. Energy*, 215, 338–47, 2018.
88. Venkatesan, S.V., Dutta, M., Kjeang, E., Mesoscopic degradation effects of voltage cycled cathode catalyst layers in polymer electrolyte fuel cells. *Electrochem. Commun.*, 72, 15–8, 2016.
89. Katayanagi, Y., Shimizu, T., Hashimasa, Y., Matsushita, N., Yamazaki, Y., Yamaguchi, T., Cross-sectional observation of nanostructured catalyst layer of polymer electrolyte fuel cell using FIB/SEM. *J. Power Sources*, 280, 210–6, 2015.
90. Tian, T., Tang, J., Guo, W., Pan, M., Accelerated life-time test of MEA durability under vehicle operating conditions in PEM fuel cell. *Front. Energy*, 11, 326–33, 2017.
91. Yuan, X.-Z., Li, H., Zhang, S., Martin, J., Wang, H., A review of polymer electrolyte membrane fuel cell durability test protocols. *J. Power Sources*, 196, 9107–16, 2011.
92. Lai, Y.-H., Rahmoeller, K.M., Hurst, J.H., Kukreja, R.S., Atwan, M., Maslyn, A.J. *et al.*, Accelerated stress testing of fuel cell membranes subjected to combined mechanical/chemical stressors and cerium migration. *J. Electrochem. Soc.*, 165, F3217, 2018.
93. Bloom, I., Walker, L.K., Basco, J.K., Malkow, T., Saturnio, A., De Marco, G. *et al.*, A comparison of fuel cell testing protocols–a case study: Protocols used by the US Department of Energy, European Union, international electrotechnical commission/fuel cell testing and standardization network, and fuel cell technical team. *J. Power Sources*, 243, 451–7, 2013.
94. Zhao, J. and Li, X., A review of polymer electrolyte membrane fuel cell durability for vehicular applications: Degradation modes and experimental techniques. *Energy Convers. Manage.*, 199, 112022, 2019, https://doi.org/10.1016/j.enconman.2019.112022.

95. Kundu, S., Fowler, M., Simon, L.C., Abouatallah, R., Reversible and irreversible degradation in fuel cells during open circuit voltage durability testing. *J. Power Sources*, 182, 254–8, 2008.
96. Gazdzick, P., Mitzel, J., Sanchez, D.G., Schulze, M., Friedrich, K.A., Evaluation of reversible and irreversible degradation rates of polymer electrolyte membrane fuel cells tested in automotive conditions. *J. Power Sources*, 327, 86–95, 2016.
97. Zhang, S., Yuan, X.-Z., Hin, J.N.C., Wang, H., Wu, J., Friedrich, K.A. *et al.*, Effects of open-circuit operation on membrane and catalyst layer degradation in proton exchange membrane fuel cells. *J. Power Sources*, 195, 1142–8, 2010.
98. Zago, M., Baricci, A., Bisello, A., Jahnke, T., Yu, H., Maric, R. *et al.*, Experimental analysis of recoverable performance loss induced by platinum oxide formation at the polymer electrolyte membrane fuel cell cathode. *J. Power Sources*, 455, 227990, 2020.
99. Jouin, M., Gouriveau, R., Hissel, D., Péra, M.-C., Zerhouni, N., Degradations analysis and aging modeling for health assessment and prognostics of PEMFC. *Reliab. Eng. Syst. Saf.*, 148, 78–95, 2016.
100. Zhang, X., Guo, L., Liu, H., Recovery mechanisms in proton exchange membrane fuel cells after accelerated stress tests. *J. Power Sources*, 296, 327–34, 2015.
101. Pei, P., Chang, Q., Tang, T., A quick evaluating method for automotive fuel cell lifetime. *Int. J. Hydrogen Energy*, 33, 3829–36, 2008, https://doi.org/10.1016/j.ijhydene.2008.04.048.
102. Zhang, T., Wang, P., Chen, H., Pei, P., A review of automotive proton exchange membrane fuel cell degradation under start-stop operating condition. *Appl. Energy*, 223, 249–62, 2018, https://doi.org/10.1016/j.apenergy.2018.04.049.
103. Reiser, C.A., Bregoli, L., Patterson, T.W., Jung, S.Y., Yang, J.D., Perry, M.L. *et al.*, A reverse-current decay mechanism for fuel cells. *Electrochem. Solid State Lett.*, 8, A273, 2005.
104. Kim, J., Lee, J., Tak, Y., Relationship between carbon corrosion and positive electrode potential in a proton-exchange membrane fuel cell during start/stop operation. *J. Power Sources*, 192, 674–8, 2009.
105. Ettingshausen, F., Kleemann, J., Marcu, A., Toth, G., Fuess, H., Roth, C., Dissolution and migration of platinum in PEMFCs investigated for start/stop cycling and high potential degradation. *Fuel Cells*, 11, 238–45, 2011.
106. Shao, Y., Yin, G., Gao, Y., Understanding and approaches for the durability issues of Pt-based catalysts for PEM fuel cell. *J. Power Sources*, 171, 558–66, 2007.
107. Yamashita, Y., Itami, S., Takano, J., Kakinuma, K., Uchida, H., Watanabe, M. *et al.*, Degradation mechanisms of carbon supports under hydrogen passivation startup and shutdown process for PEFCs. *J. Electrochem. Soc.*, 164, F181, 2017.

108. Paul, T.Y., Liu, Z., Makharia, R., Investigation of carbon corrosion behavior and kinetics in proton exchange membrane fuel cell cathode electrodes. *J. Electrochem. Soc.*, 160, F645, 2013.
109. Ishigami, Y., Takada, K., Yano, H., Inukai, J., Uchida, M., Nagumo, Y. *et al.*, Corrosion of carbon supports at cathode during hydrogen/air replacement at anode studied by visualization of oxygen partial pressures in a PEFC—Start-up/shut-down simulation. *J. Power Sources*, 196, 3003–8, 2011.
110. Babu, S.K., Spernjak, D., Dillet, J., Lamibrac, A., Maranzana, G., Didierjean, S. *et al.*, Spatially resolved degradation during startup and shutdown in polymer electrolyte membrane fuel cell operation. *Appl. Energy*, 254, 113659, 2019.
111. Park, Y.-C., Kakinuma, K., Uchida, M., Uchida, H., Watanabe, M., Deleterious effects of interim cyclic voltammetry on Pt/carbon black catalyst degradation during start-up/shutdown cycling evaluation. *Electrochim. Acta*, 123, 84–92, 2014.
112. Schulenburg, H., Schwanitz, B., Linse, N., Scherer, G.G., Wokaun, A., Krbanjevic, J. *et al.*, 3D imaging of catalyst support corrosion in polymer electrolyte fuel cells. *J. Phys. Chem. C*, 115, 14236–43, 2011.
113. Lin, R., Cui, X., Shan, J., Técher, L., Xiong, F., Zhang, Q., Investigating the effect of start-up and shut-down cycles on the performance of the proton exchange membrane fuel cell by segmented cell technology. *Int. J. Hydrogen Energy*, 40, 14952–62, 2015.
114. Yu, Y., Tu, Z., Zhang, H., Zhan, Z., Pan, M., Comparison of degradation behaviors for open-ended and closed proton exchange membrane fuel cells during startup and shutdown cycles. *J. Power Sources*, 196, 5077–83, 2011.
115. Qiu, Y., Zhong, H., Wang, M., Zhang, H., Effect of relative humidity cycles accompanied by intermittent start/stop switches on performance degradation of membrane electrode assembly components in proton exchange membrane fuel cells. *J. Power Sources*, 283, 171–80, 2015.
116. Moein-Jahromi, M. and Kermani, M.J., Performance prediction of PEM fuel cell cathode catalyst layer using agglomerate model. *Int. J. Hydrogen Energy*, 37, 17954–66, 2012, https://doi.org/10.1016/j.ijhydene.2012.09.120.
117. Zhang, X., Yang, Y., Guo, L., Liu, H., Effects of carbon corrosion on mass transfer losses in proton exchange membrane fuel cells. *Int. J. Hydrogen Energy*, 42, 4699–705, 2017.
118. Park, J.-H., Yim, S.-D., Kim, T., Park, S.-H., Yoon, Y.-G., Park, G.-G. *et al.*, Understanding the mechanism of membrane electrode assembly degradation by carbon corrosion by analyzing the microstructural changes in the cathode catalyst layers and polarization losses in proton exchange membrane fuel cell. *Electrochim. Acta*, 83, 294–304, 2012.
119. Miyazawa, A., Tada, E., Nishikata, A., Influence of corrosion of SS316L bipolar plate on PEFC performance. *J. Power Sources*, 231, 226–33, 2013.

120. Wu, J., Yuan, X.-Z., Martin, J.J., Wang, H., Yang, D., Qiao, J. *et al.*, Proton exchange membrane fuel cell degradation under close to open-circuit conditions: Part I: *In situ* diagnosis. *J. Power Sources*, 195, 1171–6, 2010.
121. Wong, K.H. and Kjeang, E., Mitigation of chemical membrane degradation in fuel cells: Understanding the effect of cell voltage and iron ion redox cycle. *ChemSusChem*, 8, 1072–82, 2015.
122. Zhao, M., Shi, W., Wu, B., Liu, W., Liu, J., Xing, D. *et al.*, Influence of membrane thickness on membrane degradation and platinum agglomeration under long-term open circuit voltage conditions. *Electrochim. Acta*, 153, 254–62, 2015.
123. Zhang, J., Litteer, B.A., Gu, W., Liu, H., Gasteiger, H.A., Effect of hydrogen and oxygen partial pressure on Pt precipitation within the membrane of PEMFCs. *J. Electrochem. Soc.*, 154, B1006, 2007.
124. Yoon, W. and Huang, X., Study of polymer electrolyte membrane degradation under OCV hold using bilayer MEAs. *J. Electrochem. Soc.*, 157, B599, 2010.
125. Springer, T.E., Zawodzinski, T.A., Gottesfeld, S., Polymer electrolyte fuel cell model. *J. Electrochem. Soc.*, 138, 8, 2334–42, 1991.
126. Ren, P., Pei, P., Li, Y., Wu, Z., Chen, D., Huang, S. *et al.*, Diagnosis of water failures in proton exchange membrane fuel cell with zero-phase ohmic resistance and fixed-low-frequency impedance. *Appl. Energy*, 239, 785–92, 2019.
127. Sadeghi Alavijeh, A., Goulet, M., Khorasany, R.M.H., Ghataurah, J., Lim, C., Lauritzen, M. *et al.*, Decay in mechanical properties of catalyst coated membranes subjected to combined chemical and mechanical membrane degradation. *Fuel Cells*, 15, 204–13, 2015.
128. Gummalla, M., Atrazhev, V.V., Condit, D., Cipollini, N., Madden, T., Kuzminyh, N.Y. *et al.*, Degradation of polymer-electrolyte membranes in fuel cells: II. Theoretical model. *J. Electrochem. Soc.*, 157, B1542, 2010.
129. Zatoń, M., Rozière, J., Jones, D.J., Current understanding of chemical degradation mechanisms of perfluorosulfonic acid membranes and their mitigation strategies: A review. *Sustain. Energy Fuels*, 1, 409–38, 2017.
130. Macauley, N., Watson, M., Lauritzen, M., Knights, S., Wang, G.G., Kjeang, E., Empirical membrane lifetime model for heavy duty fuel cell systems. *J. Power Sources*, 336, 240–50, 2016.
131. Futter, G.A., Latz, A., Jahnke, T., Physical modeling of chemical membrane degradation in polymer electrolyte membrane fuel cells: Influence of pressure, relative humidity and cell voltage. *J. Power Sources*, 410, 78–90, 2019.
132. Zamel, N. and Li, X., Effect of contaminants on polymer electrolyte membrane fuel cells. *Prog. Energy Combust. Sci.*, 37, 292–329, 2011.
133. Li, H., Tsay, K., Wang, H., Shen, J., Wu, S., Zhang, J. *et al.*, Durability of PEM fuel cell cathode in the presence of Fe3+ and Al3+. *J. Power Sources*, 195, 8089–93, 2010.

134. Scott, K., Microbial fuel cells: Transformation of wastes into clean energy, in: *Membranes for Clean and Renewable Power Applications*, pp. 266–300, Woodhead Publishing Limited, Cambridge, UK, 2014.
135. Ghassemzadeh, L. and Holdcroft, S., Quantifying the structural changes of perfluorosulfonated acid ionomer upon reaction with hydroxyl radicals. *J. Am. Chem. Soc.*, 135, 8181–4, 2013.
136. Ghassemzadeh, L., Peckham, T.J., Weissbach, T., Luo, X., Holdcroft, S., Selective formation of hydrogen and hydroxyl radicals by electron beam irradiation and their reactivity with perfluorosulfonated acid ionomer. *J. Am. Chem. Soc.*, 135, 15923–32, 2013.
137. Ferreira, R.B., Falcao, D.S., Pinto, A., Simulation of membrane chemical degradation in a proton exchange membrane fuel cell by computational fluid dynamics. *Int. J. Hydrogen Energy*, 46, 1106–20, 2021.
138. Singh, R., Sui, P.C., Wong, K.H., Kjeang, E., Knights, S., Djilali, N., Modeling the effect of chemical membrane degradation on PEMFC performance. *J. Electrochem. Soc.*, 165, F3328, 2018.
139. velan Venkatesan, S., Lim, C., Holdcroft, S., Kjeang, E., Progression in the morphology of fuel cell membranes upon conjoint chemical and mechanical degradation. *J. Electrochem. Soc.*, 163, F637, 2016.
140. Gubler, L., Dockheer, S.M., Koppenol, W.H., Radical (HO•, H• and HOO•) formation and ionomer degradation in polymer electrolyte fuel cells. *J. Electrochem. Soc.*, 158, B755, 2011.
141. Moein-Jahromi, M., Kermani, M.J., Movahed, S., Degradation forecast for PEMFC cathode-catalysts under cyclic loads. *J. Power Sources*, 359, 611–25, 2017, https://doi.org/10.1016/j.jpowsour.2017.05.102.
142. Virkar, A.V. and Zhou, Y., Mechanism of catalyst degradation in proton exchange membrane fuel cells. *J. Electrochem. Soc.*, 154, B540, 2007.
143. Baldan, A., Review progress in Ostwald ripening theories and their applications to nickel-base superalloys Part I: Ostwald ripening theories. *J. Mater. Sci.*, 37, 2171–202, 2002.
144. Yao, J.H., Elder, K.R., Guo, H., Grant, M., Theory and simulation of Ostwald ripening. *Phys. Rev. B*, 47, 14110, 1993.
145. Gilbert, J.A., Kariuki, N.N., Wang, X., Kropf, A.J., Yu, K., Groom, D.J. *et al.*, Pt catalyst degradation in aqueous and fuel cell environments studied via in-operando anomalous small-angle X-ray scattering. *Electrochim. Acta*, 173, 223–34, 2015.
146. Lifshitz, I.M. and Slyozov, V.V., The kinetics of precipitation from supersaturated solid solutions. *J. Phys. Chem. Solids*, 19, 35–50, 1961.
147. Wagner, C., Theory of ageing precipitation by umlosen (annual maturation). *J. Electrochem. Rep. Bunsenges Phys. Chem.*, 65, 581, 1961.
148. Kawano, A. and Imabayashi, S., Influence of oxygen atmosphere on dissolution of platinum under potential cycling. *J. Electrochem. Soc.*, 161, F67, 2013.
149. Mayrhofer, K.J.J., Ashton, S.J., Meier, J.C., Wiberg, G.K.H., Hanzlik, M., Arenz, M., Non-destructive transmission electron microscopy study of

catalyst degradation under electrochemical treatment. *J. Power Sources*, 185, 734–9, 2008.

150. Franck-Lacaze, L., Bonnet, C., Choi, E., Moss, J., Pontvianne, S., Poirot, H. *et al.*, Ageing of PEMFC's due to operation at low current density: Investigation of oxidative degradation. *Int. J. Hydrogen Energy*, 35, 10472–81, 2010.
151. Yasuda, K., Taniguchi, A., Akita, T., Ioroi, T., Siroma, Z., Platinum dissolution and deposition in the polymer electrolyte membrane of a PEM fuel cell as studied by potential cycling. *Phys. Chem. Chem. Phys.*, 8, 746–52, 2006.
152. Macauley, N., Wong, K.H., Watson, M., Kjeang, E., Favorable effect of *in-situ* generated platinum in the membrane on fuel cell membrane durability. *J. Power Sources*, 299, 139–48, 2015.
153. Chen, H., Zhao, X., Qu, B., Zhang, T., Pei, P., Li, C., An evaluation method of gas distribution quality in dynamic process of proton exchange membrane fuel cell. *Appl. Energy*, 232, 26–35, 2018.
154. Pei, P. and Chen, H., Main factors affecting the lifetime of Proton Exchange Membrane fuel cells in vehicle applications: A review. *Appl. Energy*, 125, 60–75, 2014.
155. García-Salaberri, P.A., Sánchez, D.G., Boillat, P., Vera, M., Friedrich, K.A., Hydration and dehydration cycles in polymer electrolyte fuel cells operated with wet anode and dry cathode feed: A neutron imaging and modeling study. *J. Power Sources*, 359, 634–55, 2017.
156. Yan, X., Hou, M., Sun, L., Liang, D., Shen, Q., Xu, H. *et al.*, AC impedance characteristics of a 2 kW PEM fuel cell stack under different operating conditions and load changes. *Int. J. Hydrogen Energy*, 32, 4358–64, 2007.
157. Hasan, M., Goshtasbi, A., Chen, J., Santare, M.H., Ersal, T., Model-based analysis of PFSA membrane mechanical response to relative humidity and load cycling in PEM fuel cells. *J. Electrochem. Soc.*, 165, F3359, 2018.
158. Mukundan, R., Baker, A.M., Kusoglu, A., Beattie, P., Knights, S., Weber, A.Z. *et al.*, Membrane accelerated stress test development for polymer electrolyte fuel cell durability validated using field and drive cycle testing. *J. Electrochem. Soc.*, 165, F3085, 2018.
159. Wang, G., Huang, F., Yu, Y., Wen, S., Tu, Z., Degradation behavior of a proton exchange membrane fuel cell stack under dynamic cycles between idling and rated condition. *Int. J. Hydrogen Energy*, 43, 4471–81, 2018, https://doi.org/10.1016/j.ijhydene.2018.01.020.
160. Kang, J. and Kim, J., Membrane electrode assembly degradation by dry/wet gas on a PEM fuel cell. *Int. J. Hydrogen Energy*, 35, 13125–30, 2010.
161. Liu, M., Wang, C., Zhang, J., Wang, J., Hou, Z., Mao, Z., Diagnosis of membrane electrode assembly degradation with drive cycle test technique. *Int. J. Hydrogen Energy*, 39, 14370–5, 2014.
162. Alavijeh, A.S., Khorasany, R.M.H., Nunn, Z., Habisch, A., Lauritzen, M., Rogers, E. *et al.*, Microstructural and mechanical characterization of catalyst

coated membranes subjected to in situ hygrothermal fatigue. *J. Electrochem. Soc.*, 162, F1461, 2015.
163. Yin, Y., Li, R., Bai, F., Zhu, W., Qin, Y., Chang, Y. *et al.*, Ionomer migration within PEMFC catalyst layers induced by humidity changes. *Electrochem. Commun.*, 109, 106590, 2019.
164. Aindow, T.T. and O'Neill, J., Use of mechanical tests to predict durability of polymer fuel cell membranes under humidity cycling. *J. Power Sources*, 196, 3851–4, 2011.
165. Shan, J., Lin, R., Xia, S., Liu, D., Zhang, Q., Local resolved investigation of PEMFC performance degradation mechanism during dynamic driving cycle. *Int. J. Hydrogen Energy*, 41, 4239–50, 2016.
166. Zhao, J., Shahgaldi, S., Li, X., Liu, Z.S., Experimental observations of microstructure changes in the catalyst layers of proton exchange membrane fuel cells under wet-dry cycles. *J. Electrochem. Soc.*, 165, F3337, 2018.
167. Chang, Y., Liu, J., Li, R., Zhao, J., Qin, Y., Zhang, J. *et al.*, Effect of humidity and thermal cycling on the catalyst layer structural changes in polymer electrolyte membrane fuel cells. *Energy Convers. Manage.*, 189, 24–32, 2019, https://doi.org/10.1016/j.enconman.2019.03.066.
168. Chen, H., Zhao, X., Zhang, T., Pei, P., The reactant starvation of the proton exchange membrane fuel cells for vehicular applications: A review. *Energy Convers. Manage.*, 182, 282–98, 2019, https://doi.org/10.1016/j.enconman.2018.12.049.
169. Jahromi, M.M. and Heidary, H., Automotive applications of PEM technology. *PEM Fuel Cells*, pp. 347–405, 2022.
170. Ebrahimi, S., DeVaal, J., Narimani, M., Vijayaraghavan, K., Transient model of oxygen-starved proton exchange membrane fuel cell for predicting voltages and hydrogen emissions. *Int. J. Hydrogen Energy*, 42, 21177–90, 2017.
171. Bodner, M., Schenk, A., Salaberger, D., Rami, M., Hochenauer, C., Hacker, V., Air starvation induced degradation in polymer electrolyte fuel cells. *Fuel Cells*, 17, 18–26, 2017.
172. Dou, M., Hou, M., Liang, D., Shen, Q., Zhang, H., Lu, W. *et al.*, Behaviors of proton exchange membrane fuel cells under oxidant starvation. *J. Power Sources*, 196, 2759–62, 2011.
173. Taniguchi, A., Akita, T., Yasuda, K., Miyazaki, Y., Analysis of degradation in PEMFC caused by cell reversal during air starvation. *Int. J. Hydrogen Energy*, 33, 2323–9, 2008.
174. Wang, B., Wu, K., Xi, F., Xuan, J., Xie, X., Wang, X. *et al.*, Numerical analysis of operating conditions effects on PEMFC with anode recirculation. *Energy*, 173, 844–56, 2019.
175. Patterson, T.W. and Darling, R.M., Damage to the cathode catalyst of a PEM fuel cell caused by localized fuel starvation. *Electrochem. Solid State Lett.*, 9, A183, 2006.
176. Qin, C., Wang, J., Yang, D., Li, B., Zhang, C., Proton exchange membrane fuel cell reversal: A review. *Catalysts*, 6, 197, 2016.

177. Liang, D., Shen, Q., Hou, M., Shao, Z., Yi, B., Study of the cell reversal process of large area proton exchange membrane fuel cells under fuel starvation. *J. Power Sources*, 194, 847–53, 2009, https://doi.org/10.1016/j.jpowsour.2009.06.059.
178. Messing, M. and Kjeang, E., Empirical modeling of cathode electrode durability in polymer electrolyte fuel cells. *J. Power Sources*, 451, 227750, 2020, https://doi.org/10.1016/j.jpowsour.2020.227750.
179. Padgett, E., Yarlagadda, V., Holtz, M.E., Ko, M., Levin, B.D.A., Kukreja, R.S. *et al.*, Mitigation of PEM fuel cell catalyst degradation with porous carbon supports. *J. Electrochem. Soc.*, 166, F198–207, 2019, https://doi.org/10.1149/2.0371904jes.
180. Liu, H., Chen, J., Hissel, D., Hou, M., Shao, Z., A multi-scale hybrid degradation index for proton exchange membrane fuel cells. *J. Power Sources*, 437, 226916, 2019, https://doi.org/10.1016/j.jpowsour.2019.226916.
181. Kregar, A., Tavčar, G., Kravos, A., Katrašnik, T., Predictive system-level modeling framework for transient operation and cathode platinum degradation of high temperature proton exchange membrane fuel cells. *Appl. Energy*, 263, 114547, 2020, https://doi.org/10.1016/j.apenergy.2020.114547.
182. Baricci, A., Bonanomi, M., Yu, H., Guetaz, L., Maric, R., Casalegno, A., Modelling analysis of low platinum polymer fuel cell degradation under voltage cycling: Gradient catalyst layers with improved durability. *J. Power Sources*, 405, 89–100, 2018, https://doi.org/10.1016/j.jpowsour.2018.09.092.
183. Kinoshita, K., Lundquist, J.T., Stonehart, P., Potential cycling effects on platinum electrocatalyst surfaces. *J. Electroanal. Chem. Interfacial Electrochem.*, 48, 157–66, 1973.
184. Patterson, T., Fuel cell technology topical conference proceedings. *AIChE Spring Natl Meet*, AIChE, New York, p. 313, 2002.
185. Bi, W. and Fuller, T.F., Temperature effects on PEM fuel cells Pt/C catalyst degradation. *J. Electrochem. Soc.*, 155, B215–21, 2008, https://doi.org/10.1149/1.2819680.
186. Rice, C.A., Urchaga, P., Pistono, A.O., McFerrin, B.W., McComb, B.T., Hu, J., Platinum dissolution in fuel cell electrodes: Enhanced degradation from surface area assessment in automotive accelerated stress tests. *J. Electrochem. Soc.*, 162, F1175–80, 2015, https://doi.org/10.1149/2.0371510jes.
187. Hiraoka, F., Matsuzawa, K., Mitsushima, S., Degradation of Pt/C under various potential cycling patterns. *Electrocatalysis*, 4, 10–6, 2013.
188. Debe, M.K., Schmoeckel, A.K., Vernstrom, G.D., Atanasoski, R., High voltage stability of nanostructured thin film catalysts for PEM fuel cells. *J. Power Sources*, 161, 1002–11, 2006.
189. Kneer, A. and Wagner, N., A semi-empirical catalyst degradation model based on voltage cycling under automotive operating conditions in PEM fuel cells. *J. Electrochem. Soc.*, 166, F120–7, 2019, https://doi.org/10.1149/2.0641902jes.
190. Kneer, A., Jankovic, J., Susac, D., Putz, A., Wagner, N., Sabharwal, M. *et al.*, Correlation of changes in electrochemical and structural parameters due to

voltage cycling induced degradation in PEM fuel cells. *J. Electrochem. Soc.*, 165, F3241–50, 2018, https://doi.org/10.1149/2.0271806jes.

191. Kneer, A., Wagner, N., Sadeler, C., Scherzer, A.-C., Gerteisen, D., Effect of dwell time and scan rate during voltage cycling on catalyst degradation in PEM fuel cells. *J. Electrochem. Soc.*, 165, F805–12, 2018, https://doi.org/10.1149/2.0651810jes.
192. Takei, C., Kakinuma, K., Kawashima, K., Tashiro, K., Watanabe, M., Uchida, M., Load cycle durability of a graphitized carbon black-supported platinum catalyst in polymer electrolyte fuel cell cathodes. *J. Power Sources*, 324, 729–37, 2016, https://doi.org/10.1016/j.jpowsour.2016.05.117.
193. Garcia-Sanchez, D., Morawietz, T., Da Rocha, P.G., Hiesgen, R., Gazdzicki, P., Friedrich, K.A., Local impact of load cycling on degradation in polymer electrolyte fuel cells. *Appl. Energy*, 259, 114210, 2020, https://doi.org/10.1016/j.apenergy.2019.114210.
194. Ahmadi Sarbast, V., Modeling of proton exchange membrane fuel cell performance degradation and operation life, MSc thesis, 2021.
195. Li, Y., Pei, P., Ma, Z., Ren, P., Wu, Z., Chen, D. *et al.*, Characteristic analysis in lowering current density based on pressure drop for avoiding flooding in proton exchange membrane fuel cell. *Appl. Energy*, 248, 321–9, 2019.
196. Dunsmore, L., Uddin, A., Zhang, H., Wu, G., Litster, S., Non planar platinum group metal-free fuel cell cathodes for enhanced oxygen transport and water rejection. *J. Power Sources*, 506, 230188, 2021.
197. Yang, Z., Jiao, K., Liu, Z., Yin, Y., Du, Q., Investigation of performance heterogeneity of PEMFC stack based on 1+ 1D and flow distribution models. *Energy Convers. Manage.*, 207, 112502, 2020.
198. Hu, X., Song, K., Niu, W., Zhang, T., Powertrain system durability in proton exchange membrane fuel cell electric vehicles: A review. *SAE Tech Pap*, 2018, https://doi.org/10.4271/2018-01-1303.
199. Fu, Z., Zhu, L., Tao, F., Si, P., Sun, L., Optimization based energy management strategy for fuel cell/battery/ultracapacitor hybrid vehicle considering fuel economy and fuel cell lifespan. *Int. J. Hydrogen Energy*, 45, 8875–86, 2020, https://doi.org/10.1016/j.ijhydene.2020.01.017.
200. Wang, Y., Moura, S.J., Advani, S.G., Prasad, A.K., Optimization of powerplant component size on board a fuel cell/battery hybrid bus for fuel economy and system durability. *Int. J. Hydrogen Energy*, 44, 18283–92, 2019, https://doi.org/10.1016/j.ijhydene.2019.05.160.
201. Zhang, H., Li, X., Liu, X., Yan, J., Enhancing fuel cell durability for fuel cell plug-in hybrid electric vehicles through strategic power management. *Appl. Energy*, 241, 483–90, 2019, https://doi.org/10.1016/j.apenergy.2019.02.040.
202. Marx, N., Hissel, D., Gustin, F., Boulon, L., Agbossou, K., On the sizing and energy management of an hybrid multistack fuel cell – battery system for automotive applications. *Int. J. Hydrogen Energy*, 42, 1518–26, 2017, https://doi.org/10.1016/j.ijhydene.2016.06.111.

203. Selvaganesh, S.V., Selvarani, G., Sridhar, P., Pitchumani, S., Shukla, A.K., Durable electrocatalytic-activity of Pt–Au/C cathode in PEMFCs. *Phys. Chem. Chem. Phys.*, 13, 12623–34, 2011.
204. Selvaganesh, S.V., Selvarani, G., Sridhar, P., Pitchumani, S., Shukla, A.K., A durable RuO2-carbon-supported Pt catalyst for PEFCs: A cause and effect study. *J. Electrochem. Soc.*, 159, B463, 2012.
205. Yu, P., Pemberton, M., Plasse, P., PtCo/C cathode catalyst for improved durability in PEMFCs. *J. Power Sources*, 144, 11–20, 2005.
206. Jung, W.S. and Popov, B.N., New method to synthesize highly active and durable chemically ordered fct-PtCo cathode catalyst for PEMFCs. *ACS Appl. Mater. Interfaces*, 9, 23679–86, 2017.
207. Souza, N.E., Bott-Neto, J.L., Rocha, T.A., da Silva, G.C., Teixeira-Neto, E., Gonzalez, E.R. *et al.*, Support modification in Pt/C electrocatalysts for durability increase: A degradation study assisted by identical location transmission electron microscopy. *Electrochim. Acta*, 265, 523–31, 2018.
208. Jung, W.S. and Popov, B.N., Effect of pretreatment on durability of fct-structured Pt-based alloy catalyst for the oxygen reduction reaction under operating conditions in polymer electrolyte membrane fuel cells. *ACS Sustainable Chem. Eng.*, 5, 9809–17, 2017.
209. Novikova, K., Kuriganova, A., Leontyev, I., Gerasimova, E., Maslova, O., Rakhmatullin, A. *et al.*, Influence of carbon support on catalytic layer performance of proton exchange membrane fuel cells. *Electrocatalysis*, 9, 22–30, 2018.
210. Kermani, M.J., Moein-Jahromi, M., Hasheminasab, M.R., Ebrahimi, F., Wei, L., Guo, J., Jiang, F.M., Application of foam-based functionally graded porous material flow-distributor to PEM fuel cells. *Energy*, 254, 124230, 2022.
211. Moein-Jahromi, M., Movahed, S., Kermani, M.J., Numerical study of the cathode electrode in the microfluidic fuel cell using agglomerate model. *J. Power Sources*, 277, 180–92, 2015.

Index